Miscellaneous Disbanded Virginia Light Artillery

The Virginia Regimental Histories Series

1st Edition

This series is dedicated to the State of Virginia and all of her people who lived during the War Between the States. It is the purpose of this series to preserve, as a part of our heritage, the deeds and sacrifices of these Virginians. Your support of this project is greatly appreciated.

Number 113 of 1,000

Robert H. Moore, II

Manufactured in the United States by
H. E. Howard, Inc., Lynchburg, Virginia

Printed by H. E. Howard, Inc.

ISBN-1-56190-105-9

Miscellaneous Disbanded Virginia Light Artillery

1st Edition

Robert H. Moore, II

Table of Contents

Acknowledgments 1

Preface 2

Introduction 6

Technical Background 7

1st Regiment Virginia Artillery 9
- The Beauregard Rifles 10
- The Washington/Hampton, Henrico, Peninsula and Williamsburg/Lee Artillery 11

The Wise Artillery 29

The Middlesex Artillery 40

Hanover Light Artillery 45

The Magruder Light Artillery 54

The Manchester Artillery 56

The Campbell Long Island Artillery 59

Appendix 63

Statistical Summary of Strengths and Losses 70

Rosters 79
- 1st Virginia Artillery 82
- The Wise Artillery 97
- The Middlesex Artillery 108
- The Hanover Light Artillery 117
- The Magruder Light Artilery 125
- The Manchester Artillery 129
- The Campbell Long Island Artillery 133

Bibliography 139

DEDICATION

To the memory of the men who served in the disbanded batteries - that their sacrifices, devotion to their respective short-lived batteries and service should not be forgotten.

Acknowledgments

First and foremost I want to acknowledge my wife Danette for her support and understanding throughout this project. I thank her and am grateful for her love and patience.

Harold Howard has been more than patient in waiting for some of my projects. Harold is to be ever commended for being the backbone of this series.

Much thanks is due to Mr. Jeff Weaver of Arlington, Virginia, for contributing a great deal of material and doing much legwork on the rosters of some of the batteries. Likewise, by the time I began this volume, Mr. Weaver had already started some of the narrative work for a couple of the batteries. Mr. Robert K. Krick of Fredericksburg, as ever, deserves thanks for use of his files on Virginia units. Ms. Susan Crites of Martinsburg, West Virginia also kindly provided much biographical information on men in the Wise Artillery.

Finally thanks to God for having seen me through the rocky beginnings in the last phases of this project on the horse artillery batteries and for continuing to see me through the projects to come.

Robert H. Moore, II
Kingsland, Georgia
July, 1997

Preface

Introduction to the Virginia Batteries Disbanded in the Fall of 1862

From the outset of the American Civil War, a number of artillery batteries had been organized that either were poorly equipped from the beginning or simply could not maintain their numbers, horses, and equipment due to a variety of reasons. By the late summer and early fall of 1862, the Army of Northern Virginia's desperate need for a formal reorganization of artillery was finally properly addressed. In a report by General William Nelson Pendleton dated October 2, 1862, the army's chief of artillery, then near Winchester, Virginia, presented the numerous problems in the service of the "long arm."

> It is clear that our service is now encumbered by too many artillery companies, of which some have never been strong enough, some are commanded by inadequate officers, and some, though well officered and entitled to honor for excellent service, are so reduced in men and horses as scarcely to leave room for a hope of their restoration to efficiency. It becomes, therefore, an imperative duty to relieve the service and the Confederacy of this burden, so expensive every way, but especially in its enormous consumption of horses, if such relief can be legally and justly effected. The only practicable mode of accomplishing the object is to determine by the fairest standard that can be selected what companies should be dispensed with, to relieve from duty the officers of such, and to assign the men and equipments to other companies retained. In endeavoring to reach right conclusions in the premises, I have sought with great care the actual merit and condition of the several batteries connected with this army; and though it were vainly presumptuous to suppose that I had escaped error, I feel assured there is as little mistake as the complexity of the case and the limited time for investigation fairly admit.
>
> Having thus sought for the data on which to proceed, I found it requisite to apply to them certain principles toward reaching right judgement. These three seemed to me to cover the ground:
>
> First, services rendered; second, efficiency of officers thus and otherwise evidenced; third, existing condition and prospects of the battery.

1st. Laudable service undoubtedly entitles a company to honorable continuance, provided it be not forbidden under one or both of the other conditions.

2nd. Officers thoroughly efficient have a prior claim. Good service for a season, under special circumstances, may have been rendered where some essential requisites for maintaining a battery through protracted difficulties are lacking. To pass upon such characteristics is delicate, yet, under existing responsibilities, essential.

3rd Where the two preceding conditions concur it is probably best to invigorate a battery to the utmost practicable; but if either fails, in case of a company much reduced below the service standard, it would seem right to merge it in some others.

In the evaluation of the batteries, the "reduced below the service standard" was the most common reason for disbanding several of the batteries. Pendleton's suggestion led to Army of Northern Virginia Special Orders Number 209 dated October 4, 1862. Within the orders were given the directions by which batteries were to be disbanded and consolidated. Though batteries from a few Confederate states were involved, Virginia batteries were the most affected. The following Virginia batteries were involved (and respective books listed within the Virginia Regimental Histories Series):

From the 1st Regiment Virginia Artillery:

1) Co. C (2nd , Sands'/Ritter's Henrico Artillery (unit history included in this volume). - men assigned to Halifax Lt. Artillery & Albemarle Artillery (Young's battery yet to be published. Albemarle Artillery found in *Amherst Artillery, Albemarle Artillery, and Sturdivant's Battery* by W. Cullen Sherwood & Richard L. Nicholas)

2) Co. F, Garrett's/Coke's Lee/Williamsburg Artillery (unit history included in this volume). - men assigned to 3rd Richmond Howitzers & 2nd Richmond Howitzers (both batteries found in *The Richmond Howitzers* by Lee A. Wallace, Jr.)

3) Co. H, Wyatt's Albemarle Artillery was ordered to be disbanded, but the battery became an independent unit. (See *Amherst Artillery, Albemarle Artillery, and Sturdivant's Battery* by Sherwood and Nicholas)

Independent Virginia batteries affected:

4) Johnson's Bedford Artillery (for unit history see The Charlottesville, Lee Lynchburg, and Johnson's Bedford Artillery by Robert H. Moore, II). - men assigned to Purcell Artillery (*The Purcell, Crenshaw, and Letcher Artillery* by Peter S. Carmichael) or Blount's Lynchburg Artillery (The Richmond Fayette, Hampden, Thomas, and Blount's Lynchburg Artillery by R. H. Moore, II)

5) Chapman's Dixie Artillery (for unit history see *The Danville, Eighth Star New Market, and Dixie Artillery* by R. H. Moore, II) - men assigned to the Purcell Artillery (*The Purcell, Crenshaw, and Letcher Artillery* by P. S. Carmichael)

6) Ancell's Fluvanna Artillery (2nd Battery) (for unit history see *The Fluvanna Artillery* by David G. Martin) - men assigned to Snead's 1st Fluvanna Artillery (see above cited volume)

7) Nelson's Hanover Artillery (unit history included in this volume). - most men assigned to the Amherst (*Amherst Artillery, Albemarle Artillery, and Sturdivant's Battery* by Sherwood and Nicholas) or the Ashland Artillery (*Ashland, Bedford, and Taylor Light Artillery* by Marilyn Brewer Koleszstory included in this volume) - most men assigned to the *Orange and Morris, Orange, and King William Artillery* by Gregory J. Macaluso)

9) Rogers' Loudoun Artillery (*The Brooke, Fauquier, Loudoun, and Alexandria Artillery* by Michael J. Andrus) - men assigned to the Fauquier Artillery (see above cited volume)

10) Page's Magruder Light Artillery (unit history included in this volume) - most men assigned to Jordan's Bedford Artillery (see *Ashland, Bedford, and Taylor Light Artillery* by M. B. Koleszar)

11) Fleet's Middlesex Artillery (unit history included in this volume) - most men assigned to the Amherst (*Amherst Artillery, Albemarle Artillery, and Sturdivant's Battery* by Sherwood and Nicholas), Ashland (see Ashland... by Koleszar), or the Jackson Flying Artillery (yet to be published)

12) Anderson's Thomas Artillery (see *The Richmond Fayette, Hampden, Thomas, Virginia Artillery* by R. H. Moore, II) - most men assigned to the Hampden Artillery (see above cited volume)

13) Thompson's Portsmouth Light Artillery (see *Grimes' Battery, Grandy's Battery, and Huger's Battery Virginia Artillery* by R. Thomas Crew, Jr. & Benjamin H. Trask).

14) Leake's Turner Artillery (see Goochland Light, Goochland Turner, and Mountain Artillery by Jeffrey C. Weaver) - men assigned to the King William Artillery (See Morris, Orange, and King William Artillery by G. J. Macaluso)

15) Turner's Artillery (see *Goochland Light, Goochland Turner, and Mountain Artillery* by J. C. Weaver) - men assigned to the Orange Artillery (See Morris, Orange, and King William Artillery by G. J. Macaluso)

16) Brown's Wise Artillery (unit history included in this volume) - most men assigned to O. B. Taylor's Battery (See *Ashland, Bedford, and Taylor Light Artillery* by M. B. Koleszar)

Though a number of other batteries were disbanded before and after Special Orders Number 209, the batteries covered in this volume, apart from those already mentioned, include:

From the 1st Regiment Virginia Artillery:

1) Co. A. - Smith's/Fraser's Washington/Hampton Artillery, disbanded in June 1862 and the men reassigned to Co. F, Garrett's/Coke's Lee/Williamsburg Artillery.

2) Co. C (1st) - Schaeffer's Beauregard Rifles, mustered out per S.O. Number 71, Headquarters, Army of the Potomac (C.S.), 11/13/61 (not to be confr's Beauregard Rifles of Lynchburg).

3) Cosnahan's/Jones' Peninsula Artillery, disbanded June 1862 and men reassigned to other batteries in the regiment, while a few went to Page's Magruder Light Artillery.

And the independent battery of

4) Weisiger's Manchester Artillery - disbanded per S.O. Number 92, A & IGO, 4/15/63, and men reassigned to the Courtney Artillery (not yet published). However, the battery had actually been reassigned to the Courtney Artillery since Captain Weisiger had resigned 8/19/62.

Introduction

After Virginia cast her lot with the Confederate State of America on April 17, 1861, her sons went to work organizing military formations to defend her from the expected hordes of Visigoth invaders from the North, as ran the general perception at the time. The basic element of the 19th century military structure was the company. Prominent men in many communities set out to raise companies when secession was an accomplished fact. Since companies were not large, most were consolidated with other similar units to form squadrons, battalions, regiments, brigades, divisions, and corps.

The Confederate government established minimum requirements for companies which many of those trying to recruit could not attain; such companies were never recognized by the Southern government. Many of those companies have been lost to history, but others have some small bits of information about them preserved in the Official Records of the Union and Confederate Armies or in original documents filed at the National Archives in Washington or the Library of Virginia in Richmond.

Other companies were organized as the war progressed for limited duties or were just not viable units. This volume attempts to cross reference those companies which became part of some larger organization after it was formed.

This volume is organized by arm of service. Under each of the major subdivisions units are arranged alphabetically for those known by a colorful local designation or the commander's name, followed by units designated by some number. This aspect of the book duplicates some information presented in Lee A. Wallace's *Guide to Virginia Military Organizations 1861-65.*

This is necessary to ensure each unit is clearly identified. The format of this book differs from Mr. Wallace's Guide in that units identified by a colorful name or the name of the commander are listed first, followed by units designated by a number. All named units cited in the Official Records are listed, and in some cases the reader is referred to the appropriate volume of that series. In such cases, this work makes no effort to detail their rosters or history of their service. If a unit is listed in the *Official Records* or has service records available at the National Archives, Library of Virginia, or some other source not presented elsewhere in this series, this information is presented in the following pages. Reserve and militia organizations are not covered in this volume, as they will be found in separate volumes prepared for this series.

Jeffrey C. Weaver

Technical Background

Artillery equipment during the Civil War is not generally well understood. To assist the reader, the following tables give the basic equipment list and specifications for cannon used by Confederate artillerists during the War. Descriptions of what each item was used for may be found in several reference books on the subject. A basic course in the artillery's function is found in *Cannons, An Introduction to Civil War Artillery* by Dean S. Thomas. This volume is inexpensive, and has illustrations of the equipment described.

Each ideally outfitted light artillery battery had the following equipment for each gun:

sponges & rammers	2	sponge covers	2
worm & staff	1	handspikes	2
sponge bucket	1	prolonge	1
vent cover	1	tar bucket	1
water bucket	2	gunner's haversacks	2
tube pouch	2	vent pouch	1
gunner's pincers	1	tow hook	1
thumb stalls	2	priming wire	1
lanyards	2	gunner's gimlet	1
fuse cutter	1	tarpaulin	1

These items were in short supply in the Confederacy. Most were improvised by military craftsmen, captured, or imported from abroad. Any battery with a full stock of these items could consider itself lucky.

Batteries usually began the war with inferior guns, which were discarded as better cannon were captured, made, or imported. Thus a single unit might use four or five types of cannon during its history. Some of the terms may be unfamiliar to modern readers, therefore the following chart on the types of artillery is provided for reference:

Piece	Bore	Range yds.	Weight lbs.
3-inch ordnance rifle	3.00"	1830	967
6-pounder (bronze)	3.67"	1523	884
10-pounder Parrott	3.00"	1830	899
12-pounder Napoleon	4.62"	1680	227
12-pounder mtn. howitzer	4.62"	900	220
20-pounder Parrott	3.67"	2100	1750
24-pounder (bronze)	5.82"	1322	1318

Each gun typically was pulled by a team of four horses and served by a crew of seven to ten men, composed of:

In-charge	Sergeant
Gunner	Corporal
Ammunition Chest	1-3 privates
Ramrod	1 private
Loader	1 private
Aimer	1 private
Firer	1 private

One or more teamsters would be present with the horses, or in the rear with them. On occasion, especially as attrition took men from various units, guns were served with smaller crews, with some loss in speed.

Introduction to the Miscellaneous Disbanded Virginia Light Batteries of the 1st Regiment Virginia Artillery

The batteries covered in this section include companies A, C (1st), C (2nd), F, and G; the Washington/Hampton, Beauregard Rifles, Henrico, Lee, and Peninsula Artillery respectively. Since Schaeffer's Beauregard Rifles (Co. C - 1) left little record of their service and apparently never left the drill grounds, the history of the company has been separated from the other four batteries that served together. Thereafter, because the four batteries did serve together in a regimental type of environment, it is more beneficial to have combined the history of the four batteries into one comprehensive history.

The Beauregard Rifles
Company C. - (1st Organization)

The Beauregard Rifles were organized as early as late April 1861. However, most men show enlistments after May 1, 1861, at Manassas or Camp Stevens, Virginia. The new organization shared the same name with another artillery company that had formed in April under Captain Marcellus N. Moorman in Lynchburg, Virginia. However, the two companies had nothing to do with one another.

With Captain Frank B. Schaeffer at the helm, the company belonged to a battalion on July 21, 1861, and served in a limited capacity in the First Battle of Manassas. Reports place the company, acting as infantry during the engagement, between a section of H. G. Latham's Lynchburg Artillery and the Fauquier Artillery within one hundred yards and to the left of Lewis Ford. Though the company had been originally designated for artillery service, there were apparently numerous circumstances that prevented it from performing in its intended role in its first engagement and thereafter.

Assigned as Company F to the 1st Regiment Virginia Volunteers, the company served but briefly, being relieved from duty with the infantry regiment by order of General Joseph E. Johnston in September 1861. Reassigned as Company C (1st organization), 1st Regiment Virginia Artillery, the company was disbanded by order of General Joseph E. Johnston (authorized by General Thomas J. Jackson) in September 1861 and officially by Special Orders No. 71, Headquarters, Army of the Potomac, November 5, 1861, pursuant to authority from the War Department dated October 22, 1861. There was apparently no unit assignment for the remaining members of the battery after having been disbanded. By the close of the company's service, Captain Schaeffer had been detailed elsewhere, leaving drill master and young 1st Lieutenant John Pelham to handle the disbandment of the Beauregard Rifles. It is questionable if the company ever received field pieces. Furthermore, there is no evidence that the company ever left the Manassas/Centreville area.

The Washington/Hampton, Henrico, Peninsula and Williamsburg/Lee Artillery

For Virginia, the Peninsula, and Home: 1861

The 1st Regiment Virginia Artillery was organized in September 1861 from companies which had served in an unorganized battalion led by Major George Wythe Randolph. The unit was originally mustered as the 2nd Virginia Artillery, but was officially reorganized as the 1st regiment. Of the four batteries covered in this volume, only the Henrico Artillery can claim pre-regimental existence in Major Randolph's prior unorganized battalion. Formerly serving as Company A of Randolph's command, the battery was commanded by Captain Johnson H. Sands. The Washington/Hampton, Lee/Williamsburg, and Peninsula Artillery all emerged elsewhere from the 32nd Virginia Infantry before joining the 1st Regiment Virginia Artillery on September 12, 1861.

Organized as early as February 1861, the Washington/Hampton Artillery came into service with primitive field pieces. Armed with little more than two iron six-pounder guns taken from pilot boats, the battery needed better ordnance. On April 20, eight to ten citizens of Hampton, seeing the Gosport Navy Yard ablaze, boarded a pungy boat and headed for the yard to salvage what they could. After narrowly avoiding a U.S. gunboat, the crew returned successfully after their search with an unmounted 12-pounder Dahlgren boat howitzer. The boat howitzer would remain unmounted until after the company was officially mustered into service. On May 13 the men of the Washington Artillery were enlisted by Colonel Charles K. Mallory in Elizabeth City County. The following day the company was mustered into service by Colonel Benjamin S. Ewell. Forty-three-year-old Charles L. Smith captained the company. Smith's company temporarily fell under the charge of Major John B. Cary along with his own Wythe Rifles.

While Smith's Company signed the enlistment papers, the Lee/Williamsburg Artillery, made up of men from James City and York Counties, under Captain William Robertson Garrrett did the same in Williamsburg. An 1858 graduate of the College of William and Mary and the son of the mayor of Williamsburg, Garrett had been attending law school at the University of Virginia until the war broke out. For a while the Williamsburg Artillery would be a part of Colonel Ewell's small unorganized battalion.

Within ten days of its enlistment, the Washington Artillery was already in danger of being captured by Federals from Fort Monroe. From a camp on the farm of Fayette Sinclair on the Back River Road, the young battery was ordered out to a position ten miles to the west at Hood's Mill. With no horses or carts to move the pieces, the men of the Washington/Hampton Battery resorted to borrowing any animals possible from surrounding farms. Private William T. Westwood recalled that the sight was amazing:

> Two oxen at the pole of one gun and a mule in the lead; a pony and an eighteen hand horse hitched to another, and the third one with two mules and an ox hauling it. By perseverance and a liberal use of hickory switches, we succeeded in reaching Yorktown.

Making its trek on May 27, the company was worn and tired by late that night, only to find the next day some of the real rigors of life in the field as soldiers. Without much camp equipage the company found shelter, along with the Wythe Rifles, in the 18th-century Grace Episcopal Church in Yorktown. Using the long horizontal gravestones as mess tables, the company caught its wind from the long march of the previous day and enjoy a few days of light duty. The poorly equipped battery would hold little more than stationary posts until they received reliable horses, caissons and harness.

A few days before the arrival of Smith's Battery at Yorktown, Captain J. B. Cosnahan's Peninsula Artillery was enrolled in Upper Grafton in York County on May 18. It then marched to Williamsburg. Upon arriving, the new company was mustered into service by Colonel Ewell and joined many of the other companies from Ewell's battalion, including the Williamsburg Artillery at nearby Camp Page. Though the camp was intended for a camp of instruction, the artillery companies still had no guns.

By June, General John B. Magruder had beefed up the defenses on the peninsula with a total of 3,400 troops at Yorktown along with the 400 others already under Colonel Ewell. Within days, elements of Magruder's army and Federals from Fort Monroe would clash at Big Bethel. Smith's, Cosnahan's, and Garrett's companies still remained ill-equipped and stood no chance of active service as yet.

By the middle of June, the Peninsula and Lee Artillery, still serving with Ewell's battalion, were part of an organization that General Magruder began to call the Peninsula Guard, the nucleus of a regiment. By July 1, the Washington Artillery was also brought into this forming nucleus, along with seven other companies other than the Peninsula and Lee Artillery, to form the 32nd Virginia Infantry. On July 6, Colonel Ewell officially received the commission that put him in command of the new regiment with Lieutenant Colonel John B. Cary as executive officer.

Upon the organization of the regiment, the artillery companies had a below-average complement of men. The Lee Artillery had a strength of 60 men, including four officers (Captain Garrett, 1st Lieutenant James L.

Lawson, and 2nd Lieutenants Nathan Richard and John Archer Coke) and twelve non-commissioned officers. The Peninsula Artillery also had a like number of men with Captain Cosnahan, 1st Lieutenant William B. Jones, and 2nd Lieutenants Orlando F. Shield and Robert A. Bright and nine non-commissioned officers. Finally, the Washington Artillery had a total of 69 including Captain Smith, 2nd Lieutenant Jesse P. Hope, and 3rd Lieutenant Thomas Tabb.

By the time of the organization of the regiment the supply situation differed from company to company. The Washington Artillery had three mounted guns on homemade carriages with cart wheels. Still wearing dark blue uniforms issued in Hampton, the men also carried some cutlasses.

Soon after the regiment was organized, the Peninsula Artillery on July 24 received four 12-pounder field guns, four caissons and battery equipage. They also had new uniforms finally issued to them including 55 coats (or pea jackets), 55 pairs of pants, 55 pairs of drawers, 13 pairs of shoes, 110 pairs of socks, and 55 blankets. The Lee Artillery remained without guns. So did another artillery company, the James City Artillery. As a result the two continued to serve as infantry. The Lee Artillery would be paid as infantry through the August 31, 1861, muster.

By September the artillery batteries had finally reached a degree of proficiency justifying General Magruder's proposal to form a regiment of artillery. In a letter to the Governor of Virginia, Magruder wrote:

> The efficiency of the artillery will be much improved by this organization, and by sending four companies of infantry to Col. Ewell's Regt (the 32nd), the efficiency of which is now greatly impaired by its present anomalous organization of six companies of infantry and four of artillery....

On September 12, 1861, the Confederate Adjutant and Inspector General's Office officially ordered the transfer of the Washington, James City, Peninsula, and Lee batteries from the 32nd Virginia Infantry to the newly organized 1st Regiment Virginia Artillery. Among the numerous other artillery companies joined in the organization was Captain Johnson H. Sands' Henrico Artillery.

The history of Sands' Battery up until the organization of the 1st Regiment Virginia Artillery remains vague. In May 1861, the battery had been in quarters at the Baptist College in Richmond. The time spent there during the spring month had led to at least two events of note. On May 30, Private Henry Bamberry fell from the third story of a college building, fracturing his right eye and suffering a broken leg. On account of his severe injuries, the private received a medical discharge. Private William H. Garnes fell asleep while on guard duty at 10 p.m. on the same day as Bamberry's fall. He had only been with the company for less than a month. The eighteen year old was brought before a court martial on June 3 at the Hermitage Fair Grounds.

Charged with having violated the 46th Article of War, the young private pleaded guilty but claimed he had been sick for four days prior to the incident. In his statement he claimed: "the Sgt. Told me I had to go on guard, I told him I was sick but he said I had never been in guard and I had better go then. I told my whole squad that I was sick. I went on guard and was taken __ that night. I have only been mustered into service three weeks today and have never had the articles of war read to me." Despite his excuse, he was found guilty of the specification and charge and was sentenced to hard labor for 30 days in the presence of the company with a ball and chain weighing six pounds attached to his leg. Following his days of labor there is no further record of young Garnes.

The first annotation made in the company rolls beyond the May events placed the battery at Jamestown Island by August, where it had arrived from Richmond via steamer that month. The uniforms and equipment of the battery were noted as poor in quality; uniforms and outfit of company being "very bad", goods "not suitable for service in the field", and discipline and drill "not very good." From Jamestown the battery moved to Spatley's Farm by August 31, 1861, just prior to its becoming part of the 1st Artillery.

Less than a month after the organization of the new regiment, Smith's Battery was ordered to Ship Point with the rifle section of the 2nd Richmond Howitzers. Soon the other batteries were also put in motion. Cosnahan's, Garrett's, Hankins' and Sands' batteries and the rifle section of Standard's battery were all ordered to take post at Young's Mill in Warwick County and report to General Lafayette McLaws.

The Peninsula Artillery was the battery most described in its movements. On October 8, Cosnahan's men moved from their station at Tettler's Neck near Williamsburg to an encampment near Young's Mill and was placed in charge of redoubts numbers 1 & 2. The march encompassed a total of 22 miles. The muster rolls made note of the march stating that the march had been performed with "great difficulty from the badness of harness being wagon harness of the most inferior description." The company had been unable to procure artillery harness and found it impossible to advance or retreat. Despite the difficulties encountered, the discipline of the men was noted as good, but the members of the company were too few.

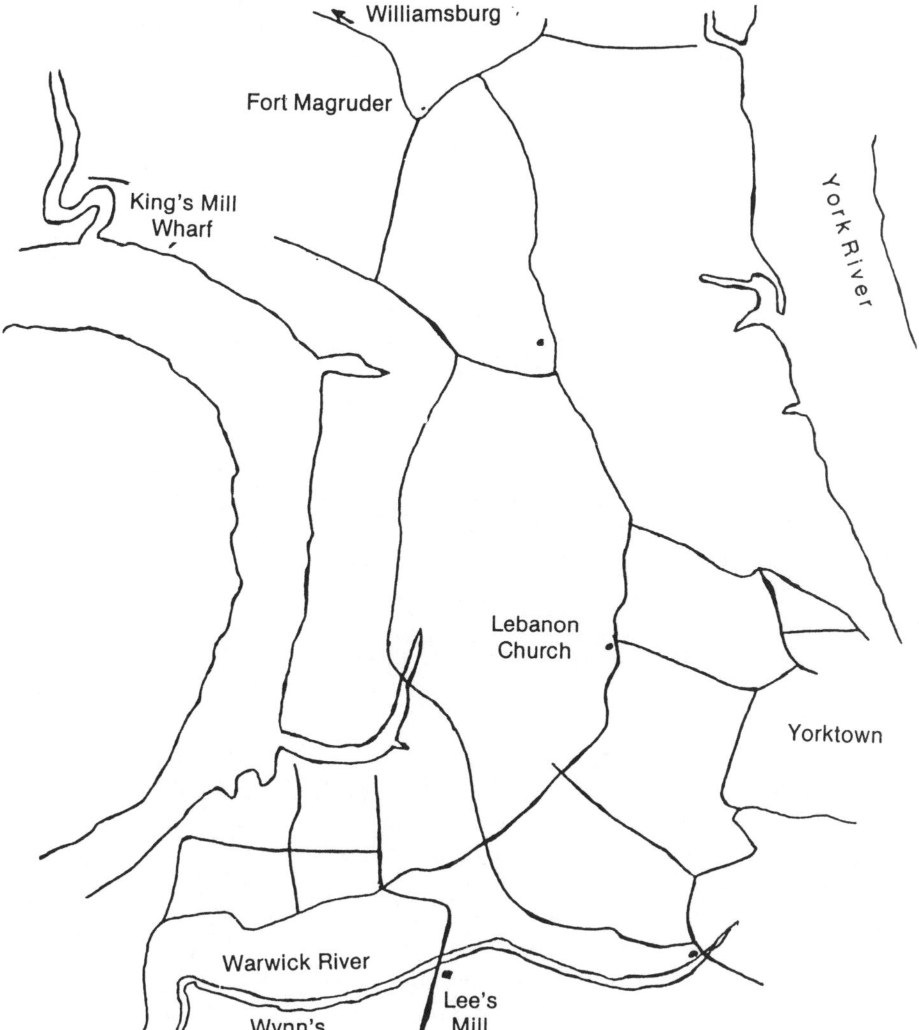
Williamsburg
Fort Magruder
King's Mill
Wharf
York River
Lebanon
Church
Yorktown
Warwick River
Lee's
Mill
Wynn's
Mill

General Magruder elaborated on the details of the batteries' new position:

> Positions for the artillery assigned to the stations of Harrod's and Young's Mills will, after consultation with General McLaws and General Rains, be designated by the chief of artillery, Colonel Randolph, and the officers and men made familiar with these positions, after which the pieces will be withdrawn and the artillery companies will be encamped---those designated for Harrod's Mill at a convenient place in rear of that station, and those intended for Young's Mill in a similar position. At each of these stations a school of practice will be established by Colonel Randolph--that on the right to be immediately commanded by Lieutenant Colonel Cabell, and that on the left by Major Brown, both of the Second Regiment Virginia Artillery, the whole under command of Colonel Randolph, of the same regiment.... The two 32-pounder guns on field carriages and the 12-pounder guns assigned to the batteries of Cosnahan's, Garrett's, and Hawkins', respectively, will be placed in position by Brigadier General McLaws in the work being erected on Young's farm. Those pieces will be served by companies to which they have been assigned.

By October 12, Magruder had become disturbed by Federal movements. Anticipating an enemy move on Lee's Store the next day, Magruder ordered McLaws to move on October 13 at early dawn with the Second Louisiana, Tenth Georgia, and the Fifteenth Virginia regiments, with Stanard's rifle section, Sands' Battery, and four companies of cavalry that were to be sent from Colonels Cobb and Johnston. After lying in ambush at Lee's Store until the 21st, the contingent was forced to move out a scouting party to better understand the Federal intentions. After a brief encounter, the demi-brigade marched to Smith's Store that afternoon to reinforce the scouting party, only to return to Lee's Store in the same evening. The following day, another force was sent forward to Smith's Store, including two guns from Sands' Battery to lay an ambush. The effort proved futile, leaving the group to return as they had the day before. By the close of the month the Peninsula, Henrico, and Lee Artillery had all assembled at Young's Mill.

While all of the other batteries had been busy in the area of Young's Mill, the Washington Artillery had encountered little more excitement. Remaining in the area of Ship Point, the company had moved once to Camp Fayetteville, seven miles, then to Bay Tree, eight miles, before returning fifteen miles to Ship Point. By the end of October, at Ship Point, the company made an account of its supplies. The battery included two iron 6-pounders with limber chests and no caissons, and one brass 12-pounder howitzer with limber chest and caisson. In addition the men of the company had 35 cutlasses as cumbersome side weapons.

Seldom was the dull camp routine disturbed for the batteries during the next month. A few probes were sent forward on occasion to test the Federals. On November 25, a section of Sands' Battery under Lieutenant Walter Robertson accompanied Captain Adams' company of the 3rd Virginia Cavalry to within three miles of Hampton. Encountering the enemy at some juncture, Robertson unleashed his guns and three shells dispersed the Federals.

As the chill of winter closed in, the Peninsula, Lee, and Henrico batteries all assembled again at Young's Mill for their first and last winter encampment over Christmas as active batteries. Disease, responsible for so many deaths in the war, was no stranger during the time in winter quarters to the batteries of the 1st Regiment. On December 26, Private John S. Morgan became the first recorded member of the Peninsula Artillery to fall prey when he died at his home nearby on the peninsula.

While December took its toll on members home on sick leave, all was not a loss during the month. By December 31, the Peninsula Battery received a late Christmas present and received a complete set of harness in good condition. The problem encountered on the march two months earlier had been remedied. The Washington/Hampton Artillery meanwhile was still stationed at Ship Point in possession of four iron 6-pounders and limbers, two 6-pounder caissons, one 12-pounder brass boat howitzer mounted on cart wheels, and 35 cutlasses.

December did not pass without fireworks. Though in no heated engagement, the Williamsburg Artillery experienced an event of internal conflict. Lieutenant James Lawson was the center of attention. As with so many others, Lawson grew ill during the miserably cold month. While ill in quarters in Williamsburg, Lawson grew well enough on December 13 to journey to the company's camp and check on the condition of things since his absence. When he returned to camp to check in, Captain Garrett ordered him to remain in camp. Claiming that he was still sick, Lawson refused and cited that he was still on a surgeon's certificate to continue rest in quarters. When Lawson could not produce the documentation, the captain again ordered him to remain. Lawson violently refused. Garrett then demanded that Lawson surrender his sword and consider himself placed under arrest. In an angry rage, Lawson then challenged Garrett and called upon the captain to "take off his straps" and fight over the matter. Garrett refused. Lawson swore that he would surrender his sword only to General McLaws or Colonel Cabell. Apparently Garrett ordered Lawson forcibly placed under arrest. Witnesses to the blatant show of disrespect were sergeants Josiah N. Wilson and Palmer.

In just over a month, Lawson was facing a general court martial for his mutinous behavior. Charged with disrespect of a commanding officer, disobedience, and conduct prejudicial to good order and discipline, Lawson was found guilty on all counts and on January 17, ordered to be dismissed from the service. Somehow, however, the lieutenant either gained favor in someone's eyes and retained his rank or began a fraud of his own design.

By March 18, 1862, Lawson had, in some manner, obtained permission slips to go to New Orleans to recruit for the company for 60 days. Lawson's official file in the Compiled Service Records, does in fact list him as a recruiting officer on April 11. In New Orleans by March 30, the lieutenant on that day wrote a letter to the Secretary of War explaining his intent to raise a battalion of 500 men from Texas. Contradictory evidence in the service record leaves this event to the reader's imagination. Lawson had apparently strayed from his original purpose of recruiting for the company. By May, Lawson had been arrested as a spy while in Goliad, Victoria County, Texas. On May 14, Lawson was in jail in Houston, still on charges of being a spy.

Lawson's trail ends in Houston, with more confusing information. In his explanation to the law agents in Texas, Lawson described his background and intentions in that state. He claimed to have graduated from the Naval Academy and been commissioned originally as a lieutenant in the United States Navy. Sent to take charge of heavy batteries at Washington, D.C., Lawson remained "until the Burnside expedition," when he decided to resign. The time frame for the resignation alone contradicts his enlistment date by almost a year. Crossing the Potomac below Romney, he made his way into Virginia and gained a commission in the service of Virginia. Ironically, there is a slip in his record that states he was a 1st lieutenant, "4th Regiment Artillery of the 9th Brigade, 4th Division Virginia Militia" as of May 11, but of 1861. Lawson's mystery concludes in Texas with no evidence of a trial or of being set free.

The First and Last Tastes of Combat: 1862

In the opening weeks of the new year, Magruder continued his face-off with the Federals on the Virginia peninsula. Matters began to grow more dangerous along the line. Many members took chances in heading home on leave. Private John Hubbard of the Hampton Artillery was captured while away on a 24 hour pass to Fox Mill.

While most batteries remained in the same positions, the Henrico Battery moved in February to a new post at Lee's Mill. When temperatures warmed into spring, the artillery in the line of defense become more important as the siege atmosphere evolved into a long series of engagements. Now under command of Colonel Henry Coalter Cabell, the 1st Virginia Artillery Regiment settled into the defenses around Yorktown. In Cabell's report dated May 10, 1862, the positions were described thus:

> Our line of defense consisted of the fortifications at Yorktown; the redoubts (Nos. 4 and 5) near Yorktown; the line of the headwaters of Warwick River and the Warwick River itself. The narrow peninsula formed by the junction of the Warwick and James Rivers was abandoned up to a point about 5 miles from the mouth of the Warwick River, and at this point (called Minor's farm) a series of redoubts, extending from the right bank of this river nearly to Mulberry Island Fort, were constructed to check any assault of the enemy upon our right flank coming up by the way of Land's End. The Warwick River had also obstructions placed in it to prevent the approach of the enemy gunboats up this river, and were further protected by our gunboat *Teazer*, which was placed near the mouth of the Warwick.

In the placement of the 1st Regiment's guns, Cabell considered the roads that made their way up the Peninsula:

> Three roads led up from the Peninsula and crossed the line of our defenses. The first on our right was the Warwick road, that crossed at Lee's Mill; the second crossed at Wynn's Mill, and the third was commanded by the redoubts (Nos. 4 and 5) near Yorktown. The crossing at Lee's Mill was naturally strong, and fortifications had been erected there and at Wynn's Mill. Below Lee's Mill the Warwick River, affected by the tides and invested by swamps on each side, formed a tolerable protection; but the marshes could be easily made passable and the river bridged. Between Lee's and Wynn's Mills an unbroken forest extended on the right bank of the stream to a distance of about 3

miles. Two additional dams were constructed -- the one (Dam No. 1) nearest to Wynn's Mill and the other Dam No. 2. A dam, called the "upper dam," was constructed in the stream above Wynn's Mill.

As the works went under construction in early April, and as a part of the line of defense, Cabell placed a 12-pounder of Cosnahan's and a Parrott of Sands' under command of Lieutenant Ritter in the extreme right redoubt at Lee's Mill, the battery under the charge of Cosnahan. Captain Sands' three remaining pieces, and Garrett's and Read's batteries, each consisting of three pieces, occupied the remaining positions around Lee's Mill.

On April 4, General George B. McClellan, in command of the Federal army knocking on the door to the peninsula, began his advance toward Magruder's lines. The afternoon of the following day, under a cloudy and rain-filled sky, McClellan's left struck the portion of the line where the Henrico Artillery stood poised. In the ensuing action, a rifled gun of Sands' Battery, along with the Richmond Fayette Artillery, opened against an enemy artillery battery. In a fight lasting from 11 a.m. until dark the battery's losses included only one sergeant's horse killed.

For the next several days the batteries of the regiment partook in numerous engagements against Federal artillery. A few batteries continued to hold their posts with little or no action. On April 5, Smith's Washington Battery, Armistead's, Richardson's, and Page's, and the remaining pieces of Captain Nelson's and Southall's batteries, were placed at redoubts number 4 and 5 in the curtain connecting these redoubts at Yorktown, and in the intermediate positions. Little is known of the Washington/Hampton battery's involvement in any action. The Henrico Artillery would not see action until April 6, when its three pieces skirmished on the left. Garrett's Williamsburg Artillery was perhaps the least exposed of all of the batteries, being placed with Young's Halifax Light Artillery in the fortifications on Mulberry Island.

Cosnahan's battery of two guns was the most mentioned in the *Official Records* of the four batteries covered. Moving up and into the works near Lee's Mill, it became a focal point for General Magruder's attentions on April 5. Magruder arrived on the center of his line, where he anticipated the enemy to make an attempt to break through. Magruder's first evidence of action at that point was: "when the booming of cannon announced the opening of the action.... I found afterward that this proceeded mostly from Cosnahan's battery of two pieces on the extreme right, replying to five or six pieces of the enemy on the opposite hill."

On April 22, Magruder once again detailed the positions from which the batteries of the 1st Regiment would continue to operate for the remainder of the month:

> I have the honor to call the attention of the commanding general to the incomplete defenses on the Curtis farm line (our right flank), connecting Skiff Creek with Lee's Mill, and to request that the particular attention of the chief engineer and his disposable force be given to perfecting them. The rifle pits already made, connecting the redoubts, are very weak, and a portion of the line so low that the trench is full of water, and the ground rising in front would shorten the fire of our sharpshooters.
>
> The woods obstruct the fire of our artillery, which would probably, if a small belt of woods were removed, command the Mulberry Island road to a long distance, firing down Curtis' field. The possible occupation of Cosnahan's battery (our advanced work) by the enemy at an early day invests this matter with the deepest interest.

At the close of the month strengths, dispositions, and armament of some of the batteries was given in the reports. Company A, the Washington Artillery, was in Yorktown in possession of three 6-pounders and a 6-pounder caisson, and one 12-pounder brass howitzer. The roster had dropped to 50 men. Company C, the Henrico Artillery (80 men), was at Lee's Mill with Brigadier General Richard Griffith's Brigade along with Cosnahan's Peninsula Artillery (with 51 men), the Richmond Howitzers, and the batteries of Manly and Read. Meanwhile the Lee Artillery, along with Young's Halifax Artillery, remained with McLaws' Brigade.

At the close of the month many things had changed for the Confederates on the peninsula, including the arrival of General Joseph E. Johnston as overall commanding officer. Realizing that McClellan was bringing up heavy guns and understanding that his line would not be able to hold against such a threat, Johnston decided to withdraw his army farther up the peninsula. On Sunday, May 4, after a terrific demonstration by the Confederate artillery, Johnston's Army of the Potomac slipped out of the Yorktown line without being detected.

In the afternoon of May 4, the Lee/Williamsburg Artillery, along with the Richmond Fayette Artillery and a section of the Richmond Howitzers under E. S. McCarthy, accompanied General Richard H. Anderson's and Pryor's brigades out of the line near Wynn's Mill. Under orders from General James Longstreet, the units moved at a steady pace to relieve the Confederates occupying the field works in front of Williamsburg.

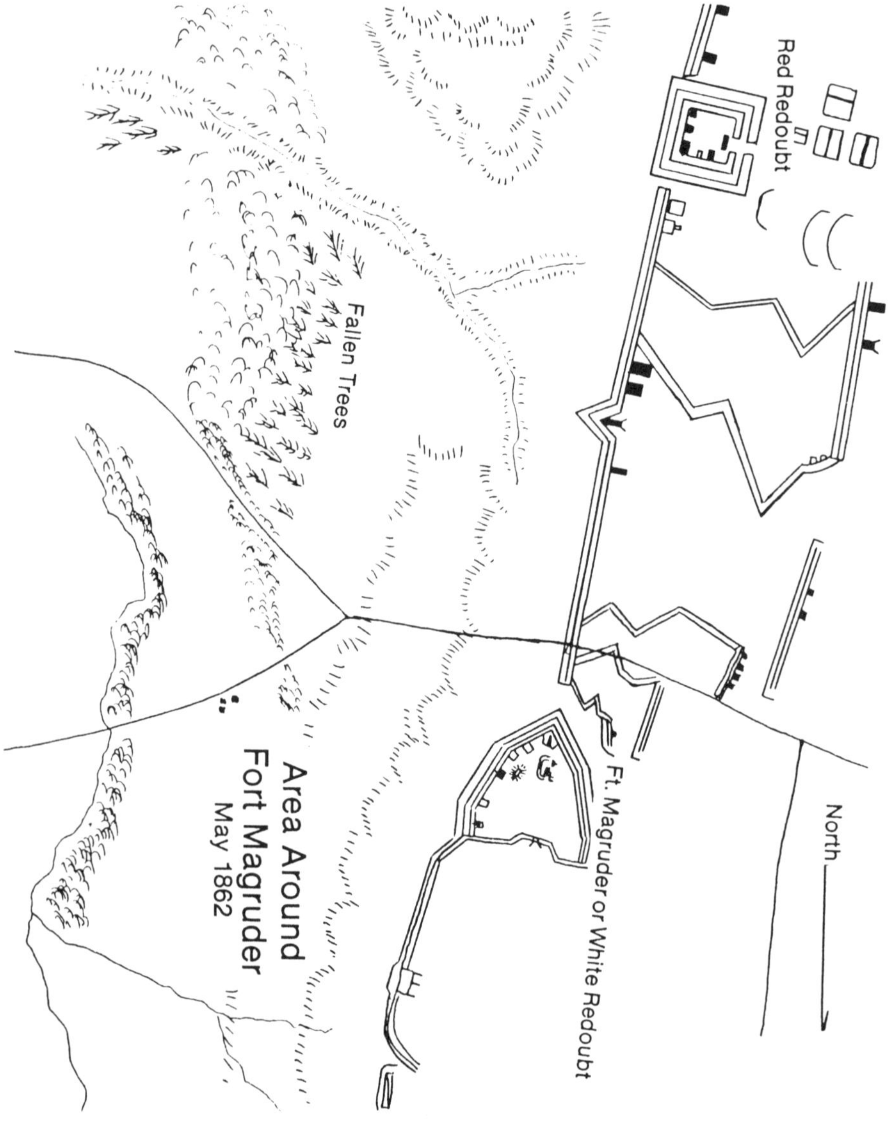

Red Redoubt
Fallen Trees
Area Around
Fort Magruder
May 1862
Ft. Magruder or White Redoubt
North

Upon arrival on the field, the brass piece from Garrett's Battery was ordered to Fort Magruder. After remaining a time in the rear of the fort, it was ordered to the front and a little to the left. In the light action that followed, a few shots were fired from guns of the 1st Richmond Howitzers, but the fire was not returned. The Williamsburg guns remained silent and were ordered again to the rear of the fort. By sunset another piece belonging to Captain Garrett's battery arrived at Fort Magruder, which was soon thereafter ordered to Redoubt number 1, on the right of the fort. Garrett's gunners had much yet to see.

At 6 a.m. on May 5, Federal sharpshooters quickly drove in Confederate pickets. Within a half hour, McCarthy's and Clopton's guns, within Fort Magruder, rang out. From the redoubt on the right, Lieutenant Coke, with an iron gun, added to the artillery punch at about 9:00 a.m. Coke's gun kept up a constant fire on the enemy for several hours before running out of ammunition. Before going silent, the crew of the iron piece believed that they had blown up a caisson in the morning. The gun was then relieved by Lieutenant Palmer's brass gun from the Williamsburg battery. Before the close of the day the crew of the brass gun claimed firing a shot that cut an enemy flag staff in half. The fire from the batteries was reported by Colonel Micah Jenkins, of the Palmetto Sharpshooters, commanding the Second Brigade: "Kept up nearlyall day with great gallantry and to good effect." Longstreet later praised the actions of the Williamsburg Artillery in the battle of Williamsburg, commenting "two pieces of Garrett's battery, under Captain Garrett and Lieutenant Coke... were distinguished for the effectiveness and rapidity of their fire." Throughout the affair, Captain Garrett commanded a brass piece in redoubt number 3.

Brigadier General J. E. B. Stuart also commented on the batteries:

> ...The enemy still did not venture to the front; but a battery of rifled pieces opened near the edge of the woods upon the open space, and, holding my cavalry near the two redoubts to the right of Fort Magruder, I assumed the direction of the pieces in those redoubts during the rest of the day, and I will here pay a merited tribute to the excellence of the execution done by them, commanded by Capt. W. Robertson Garrett, who, notwithstanding the hail-storm of bullets and shells, kept up an accurate and incessant fire upon the enemy's battery until it was silenced, and then upon his line after the brigades of infantry in the woods to the right had driven the enemy to the edge of the woods near the Telegraph road.

While no killed or wounded were recorded for the action at Williamsburg, Private Thomas Bassett of the Williamsburg Artillery was captured and sent to Fort Monroe. Furthermore Mayor Garrett, Captain Garrett's father, surrendered Williamsburg to the Federal army.

Despite the repulse of McClellan at Williamsburg, Johnston still felt threatened and continued to withdraw toward Richmond. The activities of the four batteries are not precisely known after the engagement at Williamsburg. On May 21 they were all part of Cabell's Reserve Artillery, less the howitzer companies. Apart from the Lee/Williamsburg Artillery being reported at Fairfield on May 26, the batteries do not reappear in the records until June.

With the month of June also came the disbandment of the Peninsula Artillery. On June 23, Colonel S. D. Lee expressed concern about how small Jones' Peninsula, Page's, and Young's batteries had become. The combined number of men, guns, and equipment would best be suited to reorganize as just one battery of six guns. On the day following Lee's recommendation, the Peninsula Artillery was reported to be in camp to the left of Nine Mile Road in a fort on Mrs. Price's farm. The battery had only one officer, thirty-one men, fifteen horses, and one 12-pounder gun. It was attached to Anderson's Brigade. The few men left in the company were complimented on their good efficiency. Though Lee's precise intentions were not carried out, the men of the battery were reassigned to other companies in the regiment. Since Cosnahan had died earlier in the spring, Captain William B. Jones was serving as company commander at the time of the disbandment.

In the Washington/Hampton Artillery, there had been a reorganization that brought William W. Fraser to the commanding post as captain. However, on June 14 Fraser was relieved and transferred to W. P. Carter's Battery. As was the case with the Peninsula Artillery, matters justified the disbandment of the Washington/Hampton Artillery, which was completed in June.

The number of men in the remaining batteries of the 1st Virginia Artillery continued to come under scrutiny. Also camped to the left of Nine Mile Road and on Mrs. Price's farm, the Henrico Artillery hardly had enough men to field the pieces of the battery. With one officer, 43 men, and 48 horses, the battery could not possibly bring its one Parrott and three 6-pounders into action.

On June 26, the 1st Regiment Virginia Artillery, now under Colonel John Thompson Brown, remained in reserve "near the fork of the Nine-mile road, whence they could speedily move in any direction." Colonel Brown reported of his batteries activities in late June: "My command, with the exception of the batteries, of which a more detailed account will be subsequently given, after having been in camp in its present position [Randolph's

Farm] for some days, was moved on Friday, June 27, to a point near Mrs. Price's farm, in order to be in readiness for action."

The hometown heroes from Williamsburg again gained a spotlight on June 27 and 28 when General William N. Pendleton reported of the battery: "Colonel Brown and Lieutenant Colonel Coleman had sought opportunity to be of use beyond the Chickahominy. The latter accompanied two batteries of the regiment, the Richmond Fayette Artillery, Lieutenant Clopton commanding, and the Williamsburg Artillery, Captain Coke, ordered on the morning of the 27th to report to General Lee at Mechanicsville, as he had requested. Those batteries were held as a part of the reserve of that portion of the army." Colonel Brown later reported that the two batteries were at Hogan's house in reserve during the fighting on Friday, June 27. Private Marshall Davis of the Williamsburg Artillery may have been wounded in the activities of the few days as he appears in his service record in General Hospital Number 21 in Richmond with a wound on July 2. Nothing else was listed of the injury.

The following day, Coke's Battery was moved to a point on the Darbytown road about a mile and a half above its junction with the New Market Road. There it remained without being called for until Thursday, July 3, when, in obedience to orders from General Pendleton, it returned to its old camp at Randolph's farm.

Colonel Brown elaborated on the following day's affair:

> On Saturday Colonel Coleman reported to General A. P. Hill, and in the absence of Maj. R.L. Walker, sick in Richmond, acted as chief of artillery until Wednesday, July 2, when he rejoined his regiment. The two batteries under his command from this regiment [1st] accompanied General Hill in his passage to the south side of the Chickahominy and to the vicinity of the battlefield of Monday. The Fayette Artillery was attached to Branch's brigade and the Williamsburg Artillery held in reserve...

The Williamsburg Artillery returned to its regiment on Tuesday July 1.

In closing his report of the actions of Coke's Battery, Brown commented, "The Williamsburg Artillery, being in reserve, had no opportunity to do anything amid the great superabundance of artillery and the scanty use that was made of it." Despite its relative inactivity, the Lee/Williamsburg Battery gained something from the guns captured at Frazier's Farm and Malvern Hill. Coke's Battery received a 10-pounder Parrott caisson for the battery's Parrott gun which had no caisson. As of June 28, Coke's Battery was down to 50 men present for duty. A postwar reference claimed that the battery had been presented with four captured cannon by General Longstreet during the Seven Days. With manpower so low, this claim is unlikely. The company could not have effectively manned that many guns.

The Lee Artillery was reported at Randolph's Farm performing picket duty on the Chickahominy River on July 15, then as with the regiment on July 23. Thereafter the battery does not reappear until in Camp Friend at Petersburg on August 31. The Henrico Artillery has nothing else reported of it until the disbandment in October.

The Lee/Williamsburg Artillery remained actively on the move through the end of its service. On July 12 the battery was reported as being at Blakey's Mill. Just one month later, the battery was again marching. Coke's Battery marched from Richmond to Petersburg, continuing its trek a few days later when it moved fifteen miles down the James River before returning to Petersburg and then back to Richmond on August 20.

Four days after of the battery's arrival at Richmond, Coke's men marched to Rapidan Station in Culpeper County, reaching that place on August 30. The following day, the battery moved on to Rixeyville. By some manner, Corporal Robert Caruthers found his way near the action at Sharpsburg on September 17 and was captured. He took the oath of allegiance ten days later.

When Special Orders #209 was issued, Coke's Lee/Williamsburg and Ritter's Henrico Artillery officially closed their histories. Coke received much acclaim in the order. Given credit for "service, zeal, and many excellent qualities," Coke was found without all the "requisites for efficiency." Most men from Coke's company were reassigned to the 2nd and 3rd Richmond Howitzers. When Ritter's company was disbanded it was still in Richmond or Petersburg. Being in bad health and with only one lieutenant "apparently" efficient, the men were assigned to Colonel Brown for distribution. Others entitled to special recognition and not relieved from the artillery service in the special orders included 1st Lieutenant S. H. Pendleton and 2nd Lieutenant S. H. Hawes of Coke's Battery; and 1st Lieutenant W. H. Robertson of Ritter's Battery.

Following the disbanding of the batteries, several of the men made their mark in various ways during the remainder of the war. A few stand out as of particular interest. Some men, like Captain John Coke went on to serve as staff members of other officers and followed the war to Johnston's army in North Carolina. Others such as Captain Johnson Sands fell into trouble. Sands was later mentioned on a list of prisoners from various organizations that had been previously confined in the prison who volunteered for the Wise Legion for the defense of Richmond against Sheridan's raid in 1864. As a result of his service, he was pardoned by President Jefferson Davis and released from the prison. The charge for which he had been confined is not recorded.

Others sought their fortunes in other arenas of the war. 1st Lieutenant Charles Hill, formerly of the 2nd Richmond Howitzers and the Williamsburg Artillery, rose to the ran corporal. By March 1863, Hill had followed General Magruder west and was serving as a lieutenant and A.A.I.G., District of Texas, New Mexico and Arizona. General Magruder, commanding, recommended Hill for major of artillery. In August 1863, Magruder listed Hill as major. Hill gained particular praise for his service rendered at the battle of Galveston. Hill was killed in action at Matagorda Island, Texas, about November 23, 1863.

Upon leaving the Williamsburg Artillery, Captain William Garrett accepted a commission to raise a company of cavalry for partisan ranger service in Tennessee. A postwar account described the captain's adventures:

> At this juncture the crew of the Merrimac arrived in Richmond, and about 100 of them enlisted with Captain Garrett for that service. A naval officer, having received authority from the navy to organize a fleet of small boats to attack the Galena, which was near Drewry's Bluff, offered Captain Garrett command of one division of the canoe fleet. The proposition was accepted, and the boats, manned by the Merrimac men, went rapidly down the river to Drewry's Bluff. There they reported to General Mahone, only to find that the Galena had departed, and that an order from the war department had been issued at the request of the navy department to detain all the Merrimac crew as marines to man the guns at Drewry's Bluff and for future service in the navy, the claim being made that their term of enlistment in the navy had not expired. Captain Garrett returned to Richmond, and, a few days afterward, accompanied by Maj. D. C. Douglass and others, went to Tennessee for the purpose of recruiting a battalion of partisan rangers. The party entered the Federal lines in middle Tennessee and speedily organized Douglass' battalion of partisan rangers, of which Garrett was made adjutant.

Upon the repeal of the partisan law, the battalion was mustered in as regular cavalry and consolidated with Holman's Battalion, forming the 11th Tennessee Cavalry. Ultimately becoming a part of Dibrell's Brigade of Nathan Bedford Forrest's Cavalry, it was a part of Bell's Brigade with Forrest when he surrendered at Gainesville, Alabama. Garrett had been adjutant and captain of Company B and, unlike a great many others, survived the conflict. After the war, he returned to Williamsburg to practice law but was offered a position as master of the grammar school of the college. In January 1868, he returned to Tennessee to live and married in Pulaski, Giles County. Garrett later achieved such positions as President of Giles College; superintendent of public instruction in Giles; principal of Cornersville Academy; professor of mathematics in Montgomery Bell Academy of the University of Nashville; state superintendent of public instruction; principal of the military academy at Nashville; president of the State Teacher's Association; president of the National Educational Association of the United States at the international meeting in Toronto, Canada; professor of American History at the Peabody Normal College, Nashville, Tennessee; editor of the American Historical Magazine; vice-president and chairman of the historical committee of the Tennessee association; member of the committee on history of the United Confederate Veterans; and one of the trustees of the Confederate memorial association. Living a full life, he died in Nashville, Tennessee, in 1904.

The Wise Artillery

Alburtis'/Brown's Battery

Originally formed in the fall of 1859, the Henry A. Wise Artillery was raised in Berkeley County, in what is now West Virginia, in response to John Brown's Raid on Harper's Ferry. By October 1859, the company carried 70 men on its roster. One section of the company was at Charles Town on John Brown's execution day.

The Wise Artillery was mustered into Confederate service on April 19, 1861, at Martinsburg. Originally assigned as Company B, 1st Virginia Artillery, the company was later detached from that organization as an independent battery on April 26, 1862.

The first known assignment for the Wise Artillery is found on a June 30, 1861, table of organization for Joseph E. Johnston's Army of the Shenandoah. At that time, Alburtis' four-gun battery was included in Colonel Francis S. Bartow's Second Brigade. Other units in the brigade were the 7th, 8th, and 9th Georgia Infantry, Duncan's Kentucky Battalion, and Pope's Kentucky Battalion.

While the unit was being reorganized in June, twenty-two-year-old 1st Lieutenant John Pelham was detailed as drill master. Pelham had left the United States Military Academy prior to graduation in order to offer his services to the Confederate government. When he arrived, the fuzz faced boy was probably greeted with little enthusiasm by the forty-four-year-old Ephraim Alburtis. Tactfully offering his services, Pelham soon had the company for his own to drill as he deemed necessary. In the days that followed, the inexperienced company underwent extensive training in the fields between Winchester and Martinsburg.

By late June 1861 there had been sporadic skirmishing in the mountains of western Virginia, but no general engagement. In a few short weeks however this would change. During the early days of the war, many Northerners and Southerners expected the war to be over within six months. When the Federals pushed west from their base around Washington, the Confederates were woefully unprepared to meet them. Good Confederate intelligence, however, alerted Southerners that something big was about to happen. This intelligence was relayed to the Confederate commander in Northern Virginia, General Pierre T. G. Beauregard, and to General Joseph E. Johnston in the Shenandoah Valley.

Federal commander Irvin McDowell's advance was leisurely, allowing Johnston to bring up his force from Winchester in the Valley. In preparation for the battle, Johnston and Beauregard reorganized their artillery.

Within that reorganization, the Wise Artillery and Stanard's Thomas Artillery were assigned to D. R. Jones' Brigade.

In anticipation of a move, three-days' rations were prepared on the evening of July 17. The following morning Johnston's army was put in motion eastward toward Millwood and then to Ashby's Gap in the Blue Ridge. During a halt it was announced that they were heading to Manassas Junction to aid General Beauregard in throwing back an imminent attack by the Federals. A member of the Rockbridge Artillery recalled that the news was greeted by cheers and caused the pace of the soldiers to quicken.

In the day and a half that followed, the Wise Artillery kept up with Johnston's army as it forded the Shenandoah River at Berry's Ferry, moved through Paris and the Plains on the eastern side of the Blue Ridge, and arrived at Manassas Junction in the late afternoon of July 20. Soon after arriving, Alburtis' Battery deployed with the Thomas Artillery and Jones' Brigade across McLean's Ford and into a location adequate for an "attack upon the enemy on or about the Union Mills and Centreville Road." While the guns were placed in position, nervous and green artillerists prepared for what the next day would bring.

According to Johnston's report, the first indication he had of a battle was hearing gunshots about 11 a.m. At that time he directed Colonel William N. Pendleton to hurry to the front, about four miles away, with his Rockbridge Artillery and the Thomas Artillery. The Wise Artillery, still under the watchful eye of Pelham, was to follow immediately with its guns. Arriving on Henry House Hill, Pendleton and the batteries were greeted by General Thomas J. Jackson, who directed the batteries to the crest of the ridge opposite the Henry house.

A member of the Rockbridge Artillery recollected, "We reached the top of the hill to our right in an old field and unlimbered our guns and commenced firing to the left of the direction in which we entered the field." Just to the rear of the batteries was the brigade that would gain the immortal name of the general who stood with it that day. The soon-to-be "Stonewall" Brigade lay waiting on the ground for further orders. The Wise Artillery was later placed in another position near the Robinson house.

For untried gunners, Alburtis' green artillerists and their four smoothbore guns fought hard against the enemy. The Wise Artillery is not mentioned in any casualty report, while Latham's Battery immediately adjacent, sustained only one man wounded.

General Beauregard's report praised the artillery in general. Alburtis' Battery played a key role in the Confederate victory. Beauregard wrote:

...all from the outset displaying that marvelous capacity of our people as artillerists which has made them, it would appear, at once the terror and the admiration of the enemy. As was soon apparent, the Federalists had suffered severely from our artillery and from the fire of our musketry on the right, and especially from the left flank, placed under cover, within whose galling range they had been advanced; and we are told in their official reports how regiment after regiment thrown forward to dislodge us was broken, never to recover its entire organization on that field.

Two days later, Lieutenant Pelham wrote of the battle;

...I was under a heavy fire of musketry and cannon for about seven hours, how I escaped or why I was spared a just God only knows. Rifle balls fell like hail around me. Shells bursted and scattered their fragments through my Battery - my horse was shot under me, but did not give out till the fight was almost over. I was compelled to take one of my Sergeant's horses and ride through. At one time I dismounted and directed the guns - one of the gunners asked me to dismount and shoot the Federal's flag down. I did so - you ought to have heard the cheers they gave me. I directed all my guns three or four times apiece. My men were cool and brave and made terrible havoc on the enemy. They fought better than I expected they would. The highest praise is due them. We shot down three U.S. flags and dislodged the enemy from several positions. I was complimented several times on the field of battle by general officers and a great many times after the battle was over by other officers.

You might want to know my feelings - I felt as cool and deliberate under the shower of lead and iron as if I had been at home by our fireside - I did not feel fear at any moment; I can't see how I escaped... We are battling for our rights and our homes. Ours is a just war, a holy cause. The invader must meet the fate he deserves and we must meet him as becomes us, as becomes men.

Following the fight at Manassas, the Wise Artillery disappears from reports. Like the other batteries of the Army of the Shenandoah, the Wise battery probably remained in the area of Centreville through December and went into winter quarters. Through the months that followed Manassas and into the winter of 1861-1862, the Wise Artillery experienced little change.

The staff of officers was only affected once on July 30, when 2nd Lieutenant George Murphy tendered his resignation.

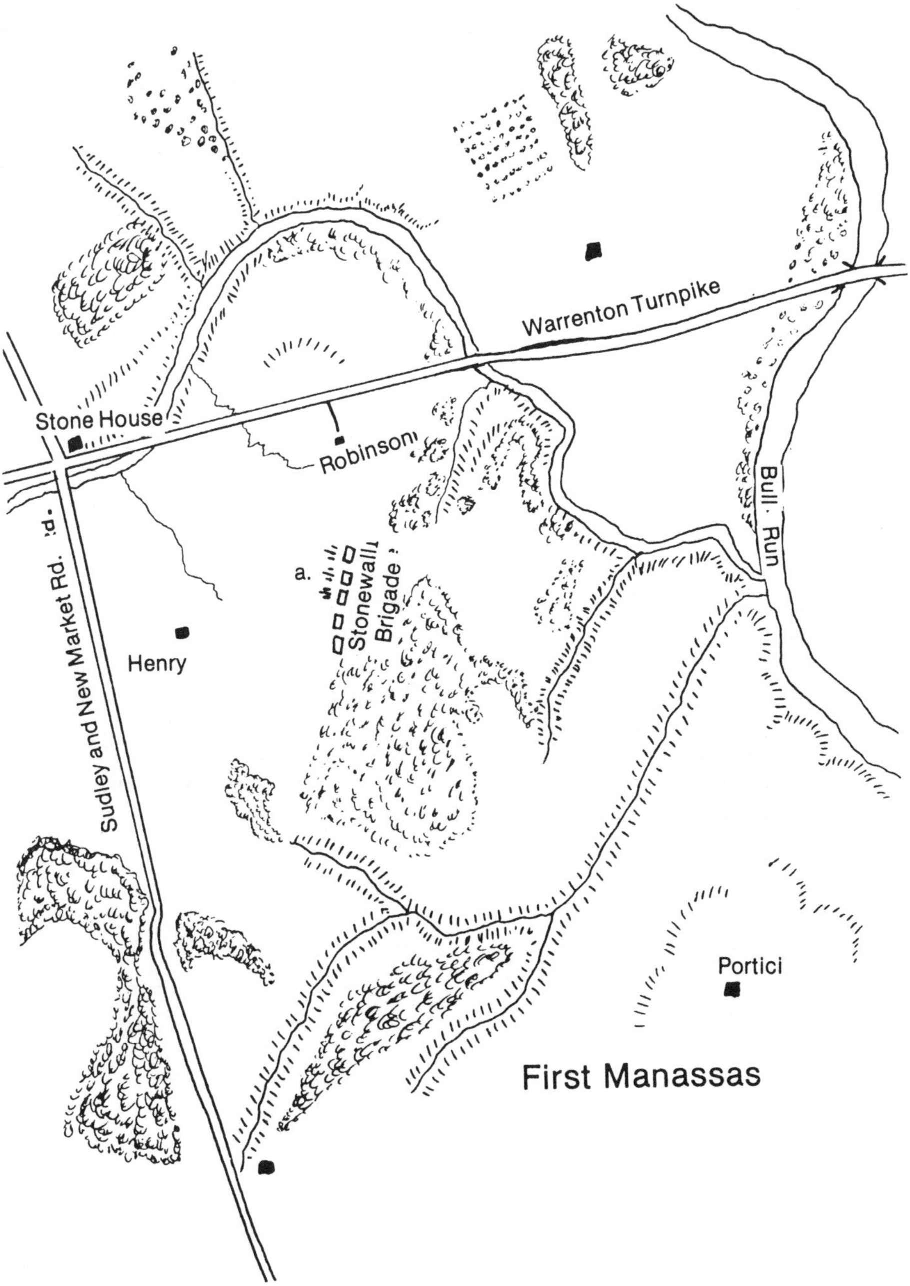

First Manassas

Winter quarters lasted through March. Despite the coming of spring, the rains proved just as miserable as winter snow. The battery experienced a great change on January 25 when Captain Alburtis tendered his resignation due to poor health. Within a week, James Stephenson Brown was promoted to the captaincy. Brown was three years older than Alburtis. Despite his age, Brown would be re-elected as captain on April 26.

While in winter quarters, the Wise Artillery became part of Colonel G. T. Anderson's Brigade of D. R. Jones' Division, along with several Georgia infantry regiments and Hart's Washington (S.C.) Battery. On February 11, Colonel Anderson made a special visit to the camp of the battery and made his first inspection. No record was made of his comments.

Other events that dominated the company's discussion included recruiting and illnesses. G. W. Hardy and Andrew Moore died of disease in March and April respectively. Enlistments helped to fill the voids from discharges and deaths. At least one member, John Patrick Mulligan, crept into the ranks of the company at age 14½.

On April 11, the Wise Artillery was camped nine miles from Louisa Court House. Covering an average of fifteen miles per day, the battery reached Richmond on April 16. Passing along Broad Street, the company had but little time in town before moving on toward New Kent Court House. The men then continued about 30 miles to Williamsburg. The Wise Artillery remained in the vicinity of Williamsburg and Yorktown for almost a month before being called upon to move into the works.

On May 2, the Wise Artillery was ordered to the fortifications at Dam Number 4. On the following day, the battery again moved, to the vicinity of Fort Magruder before being withdrawn in advance of the Battle of Williamsburg. As with many batteries, the Martinsburg battery was put to the extreme test in its flight from the peninsula, across the Chickahominy, to New Kent Court House, and to the vicinity of Richmond by the middle of May. The whole march consisted of preparations for action that never came for the Wise Artillery.

On May 23, Colonel Beverly H. Robinson of the 4th Virginia Cavalry led a column near Mechanicsville. The force included the 8th and 9th Georgia, portions of the 1st and 4th Virginia Cavalry, one 8-pounder rifled gun from the Washington Artillery, and one 6-pounder (or 12-pounder) howitzer from the Wise Artillery under Lieutenant Cyrus Kerns. The two guns were commanded by Captain Thomas L. Rosser, of the Washington Artillery. When reaching a point near Mechanicsville, enemy artillery fire was received and returned by Rosser's gun. Though the Wise gun remained silent, the wait for the following day's actions left a tense air.

The following day, between 4 and 6 a.m., the enemy advanced on the Confederate position and drove in the skirmishers. Not allowed to fire until the enemy was within one-hundred yards, the Wise gunners stood poised for action. After a rifleman of the 8th Georgia opened the affair with a shot from his gun, the enemy laid down in the wheat and opened up. The gun from the Wise battery opened fire on a regiment of Federals trying to take cover in a wood. Soon the enemy artillery joined in and commenced to cause damage. Almost immediately, Hess' saddle horse was wounded badly in the flank. While he harnessed another animal, the saddle horse of the wheel was also hit and soon after the gun was ordered to be retired by hand. After withdrawing fifty yards back to a wood themselves, the Wise gunners came under an intense artillery fire and soon were advanced upon by a great force which drove them farther to the rear. Rosser's gun accompanied the Wise piece and drew more fire from the enemy guns.

Robert Lancaster was the most severely wounded member of the Wise Artillery in the fight, having his left arm taken off near the elbow. Likewise, John Q. Herndon had his right arm taken off at the shoulder, also by a shell. Withdrawing another thousand yards to the rear, the guns were again unlimbered, but were ordered quickly to limber again and retire. Before the close of the affair, the Wise gun had withdrawn and fired from several different points, each time being forced to withdraw due to the heavy fire from enemy artillery. Of the men wounded, Herndon died the following day as a result of his wound. Ironically, his brother would die on May 25 in General Hospital Number 21 in Richmond due to rubeola.

As the waltz continued up the peninsula toward Richmond, the Wise Artillery was again reorganized in the 1st Division (General David R. Jones) as part of Major John J. Garnett's artillery battalion including Hart's Washington (S.C.) battery, Lane's (Ga.) battery, Moody's (La.) battery, and Woolfolk's Ashland Artillery. While being reorganized, the men of the Wise Artillery wanted desperately to return home and serve under Jackson in the Shenandoah Valley. Those orders never came.

Though the Wise Artillery remained around Richmond for the following month, it did little more than tend to picket duty and march around the capital periodically without seeing battle. On June 5 the Maryland men from the battery were granted their transfers to the Maryland Line to join Jackson's forces in the Shenandoah. Ironically, the Maryland men would find themselves back near Richmond within days.

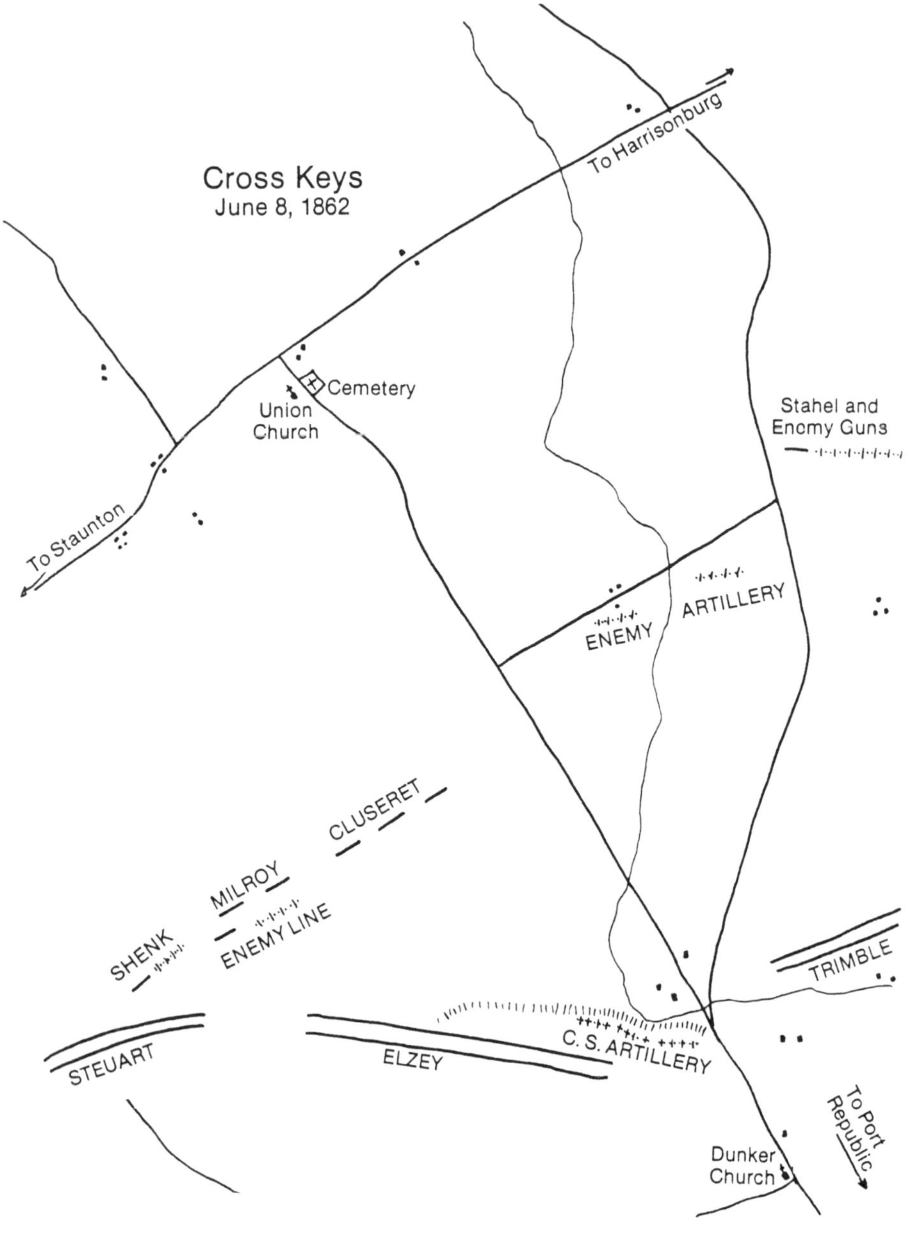
Cross Keys
June 8, 1862
To Harrisonburg
To Staunton
Cemetery
Union Church
Stahel and Enemy Guns
ENEMY
ARTILLERY
CLUSERET
MILROY
SHENK
ENEMY LINE
TRIMBLE
STEUART
ELZEY
C. S. ARTILLERY
To Port Republic
Dunker Church

On June 24 Brown's battery was reported as having 4 officers, 87 men, two 12-pounder howitzers, two 6-pounder guns, and 64 horses, Still attached to Colonel G. T. Anderson's Brigade, the battery was listed as having good efficiency. It was camped to the left of Nine Mile Road in the woods near Mrs. Price's farm.

The following day the battery again was put into a position under fire. Along with Lane's Battery, the Wise Artillery was placed in position near the Garnett's overseer's house to open against Federal working parties in opposing entrenchments. After an exchange lasting fifteen minutes, the enemy was driven off. However, Union artillery batteries soon began counterfiring with twelve rifled guns of large caliber (20 pounder Parrotts) leading to a duel that lasted for half an hour before forcing the Confederate batteries to retire.

Before noon on June 27, Brown's Battery was again called upon and directed by General Magruder to return to the same position of the day previous. Two 12-pounder howitzers, under Captain Brown, were sent forward supported by a Georgia regiment from Colonel Anderson's brigade. In position on the crest of the hill near the overseer's house on the Golding farm, the guns were placed "equidistant from Dr. and Mr. James Garnett's." At 10 o'clock Brown's two pieces opened and drove the enemy working parties from their earthworks, approximately 500 yards to the front. In turn, several long-range Federal artillery pieces opened "a terrible artillery fire from his batteries on the right of the Golding house." Meanwhile, the remainder of the Wise Artillery, consisting of two smoothbore 6-pounders, and Lane's Battery of six guns, were sent to Captain Brown's assistance. "The fire was kept up vigorously for about twenty minutes longer, when the pieces were withdrawn by Lieutenant-Colonel Lee, agreeably to instructions." The Federal guns had the upper hand in the contest once again. Before withdrawing from the field, the Wise Artillery had again suffered casualties, losing Corporal Charles W. Lucas, killed; and Privates G. W. Beard, G. T. Tinder, and Benjamin Lucas wounded. Two horses were so severely wounded that Brown was compelled to leave them on the field. One of Brown's guns was unable to participate in the affair, having been disabled by the wedging of a shell in the bore.

The following day the enemy was determined to be evacuating. Five long-range guns, directed by D. R. Jones, were ordered to open fire from the New Bridge road to enfilade the enemy position across the Chickahominy River fortified at or near the Golding house. Ordering Toombs' Brigade forward, Jones moved Brown's and Moody's batteries into the same position as the day before, near the overseer's house. Soon thereafter, Jones ordered Brown and Moody to open on enemy to the right of Toombs' position. Brown opened first, to which no fire was returned. When

Moody opened, however, fire was returned by an enfilading battery, "unmasked on the right during the previous night, and about two batteries in front." The affair lasted two hours before the batteries were again ordered to retire. The Wise Battery again suffered losses. Lieutenant Cyrus McLean Kerns was killed, Private John W. Clarke was slightly wounded and one horse killed.

Late in the afternoon of Sunday, June 29, Brown's and Hart's Washington (South Carolina) batteries were ordered to positions near the railroad, "where Captain Hart placed his two Blakely guns in position, and did handsome service until the enemy opened a plunging fire upon him from superior guns." No mention was made of the Wise battery's participation.

On Monday, June 30, the batteries moved forward with the division. They did not become engaged through the action at Malvern Hill on July 1. The following day George and William Kinzer and Charles Westphal deserted the battery, never to return to the ranks again.

Not until August 11 did the Wise Artillery break the bonds with their camp life near Richmond. Heading out along the Brook Turnpike, the battery moved with anticipation of being assigned to Jackson. On August 17 the Wise Artillery arrived within three miles of Orange Court House. The following day, the company made camp at Clark's Mountain, where it remained until the night of August 19.

In the days that followed, Brown's Battery marched across the Rapidan River, into Culpeper County, and north of Stephensburg by August 21. En route, the company passed a spy who was caught and hanged in the first tree available. The company also saw an unnamed (simply referred to as "William L-") individual that had been a member of the Wise Artillery and had deserted during the spring. After he had deserted he had gone on to serve in the 2nd Virginia Infantry, deserting it too, before joining the U.S. service. Sometime while he was acting as a picket for the Federals, he was captured near Stephensburg. When the men of Martinsburg saw the pitiful fellow, he was awaiting orders from General Longstreet to be hanged.

For the time following the passage through Stephensburg, the Wise Artillery always seemed to be a couple of steps behind the action. The line of march took them through Jeffersonton, Waterloo Bridge, Amissville, Blackwell Mills, Salem, White Plains, and finally to Thoroughfare Gap on August 28. During the passage through the gap the following day, the company came under some fire and was later ordered to remain at the gap to take up all stragglers. By Saturday, August 30, the battery had arrested more than 1,500 men.

Remaining at the gap through September 3, the Wise battery attempted to catch up with the army, passing through Hay Market, Gum Springs, and reaching Leesburg on September 5. From the river town, the battery kept pace with Longstreet's Corps across the Potomac into Montgomery County, Maryland. Brown's battery continued to serve with D. R. Jones' Division through the Maryland Campaign. In its last organization, the Wise Artillery, consisting of two 10-pounder Parrotts and one 3-inch rifle, was joined by the Fauquier, Loudoun, and Turner Artillery. Once across the Potomac, Brown's battery continued its seemingly endless trek to near Frederick, Boonesboro, Funkstown, and into familiar country at Hagerstown on September 12.

Leaving Hagerstown on September 14, the Wise Artillery found itself on contested ground the next day near Sharpsburg. In the afternoon of September 17, the Wise Artillery encountered its final action and proved its continued combat efficiency. After 4 p.m. the Federals had crossed Burnside Bridge and were pushing against Toombs' and D.R. Jones' brigades. Brown's, Richardson's, and Moody's batteries were called upon to move to the front to play against the second line of Federals advancing up the hill. After gaining permission from General Jones, Major B. W. Frobel placed Brown's battery atop the hill with the enemy 400 yards away. Brown's guns opened a "hot and well directed fire upon them, breaking their ranks and driving them back to cover of a hill from which they had just advanced." The enemy swept up the crest from the creek but were "mowed down by Brown's battery, the heroic commander of which had been wounded but a few moments before. They overcame the tough resistance offered by the feeble forces opposed to them, and gained the heights, capturing McIntosh's battery...." Brown was severally wounded by a musket ball passing entirely through his foot. Private David Blackeney was also wounded during the fighting that day.

Following the fight at Sharpsburg, the Wise Artillery became one of the most efficient batteries to go before the "chopping block" of Special Orders Number 209. The history of the Wise Artillery ended on October 4, 1862, when Generals Pendleton and Lee reorganized the artillery in the Army of Northern Virginia. While the record of the Wise Artillery was far superior to most batteries, the company had only 48 men in ranks. Captain Brown had been severely wounded at Sharpsburg and the prospects for recruiting the battery up to minimum strength were not good. The battery's men were transferred on October 8 to Colonel S. D. Lee's Artillery Battalion for reassignment. Some men found new homes in Taylor's Battery, others were assigned to the Lee Lynchburg Artillery.

The Middlesex Artillery

A Turbulent Battery 1861 - 1862

The Middlesex Artillery was organized on May 24, 1861, at Urbanna, Virginia, by Captain William C. Fleet. The battery was first assigned as Company K (1st organization) of the 47th Virginia Infantry. On August 16, 1861, the Adjutant and Inspector General's office issued Special Orders Number 252, which transferred this company to a separate battalion being formed by Major William N. Ward. It subsequently served as Company B, 55th Virginia Infantry, until April 26, 1862, when it was detached from that unit and became an independent artillery battery.

Upon separation from the 55th Infantry, Fleet's Battery became part of Colonel Reuben Lindsay Walker's Artillery Battalion. Regretfully, however, the only extant muster roll for this battery while an independent organization was prepared on April 30, 1862. At that time Fleet's company was located at Camp Barton, in Spotsylvania County. The record of events notes that the company had moved six times, covering 125 miles during the preceding two-month period. The company's enlistments were also extended for two years while in camp in Spotsylvania County and the men awarded $50 bounties.

By July 23, the Middlesex Artillery had become a part of General Ambrose P. Hill's Division of "Stonewall" Jackson's Corps. In Brigadier General C. W. Field's 1st brigade, Fleet's Battery joined the Purcell Artillery and the 40th, 47th, 55th, and 60th Virginia Infantry, and the 2nd Virginia Heavy Artillery.

On August 1, Fleet's Battery set up camp near the Purcell Artillery near the Orange and Albemarle County lines and three miles from Gordonsville. For almost a week, the battery would enjoy good weather and relaxation before moving again on August 7. On that day Hill's troops advanced through Gordonsville to within one mile of Orange Court House. Before sunrise on August 9, Hill's men were back on the Culpeper road and continued their march. Fording the Robinson River the troops were moving toward Cedar Mountain.

Before leaving camp on the 7th, the Middlesex Artillery was once again reorganized with other batteries. In Lieutenant Colonel R. L. Walker's battalion still, the battery joined with several others including the Fredericksburg, Crenshaw, Davidson's, Latham's, McIntosh's (S.C.), and Pegram's batteries.

Early in the afternoon of August 9, the booming of cannon could be heard, alerting the men to be ready for action soon. At 2 p.m. Colonel Walker received a note from Hill directing him to move his long-range guns to the front. In attempting to execute the command, the artillerists found the road impossible to advance along, being heavily clogged with wagons, ambulances, and men. Out of all of his batteries, only Pegram's and Fleet's were able to pass through due to the determination of the senior battery officers. Lieutenant Hardy would lead the Middlesex battery into action during the day.

After finally arriving on the field, Colonel Walker hastily ordered the batteries to the right of the road near the center of the Confederate line, and to the left of Jubal A. Early's brigade. While Fleet's cannoneers quickly unlimbered, the Purcell Artillery set up to their left. The batteries had been placed, unknowingly, only one hundred and fifty yards from Federal skirmishers who lay waiting in the thick green field of corn to the front.

Having just unlimbered and begun to deliver an oblique fire across their front, the batteries' position was quickly threatened when heavy firing was heard to the left. Suddenly, Confederates began pouring past, indicating that the Federals had broken through. Almost simultaneously, the Federals hidden in the corn raised up and rushed toward Fleet's and Pegram's guns. The advance through the corn took the enemy line through a brook and hindered the advance.

In his report, Colonel Stapleton Crutchfield recounted the events:

> At this moment the enemy's infantry advanced again in general line across the corn field, and Lieutenant-Colonel Walker's guns were turned on them exclusively... The enemy advancing, the rear of the guns of General Winder's division was exposed, and they were withdrawn by General Jackson's order. At the same time the enemy made a feeble effort to advance through and from the corn field, but a well-directed fire of canister from the guns of Captain Pegram and (Lieutenant) Hardy, supported by, I believe, the Thirteenth Virginia Regiment, Col. J. A. Walker, checked them, though their skirmishers got quite near under cover of the accidents of the ground. The temporary confusion on the left was soon overcome, and in a short time the enemy gave way, and our whole line advancing, the artillery moved along the road, unable to cross the brook in front through the fields. The pursuit continued until, having crossed the second brook, we came upon a large body of woods. It being deemed advisable to shell these before advancing farther, the batteries of Captain Pegram, Fleet, Braxton, and Latham were placed in position under Lieutenant-Colonel Walker some 80 or 100 yards distant, and a heavy fire opened in various directions....

In the pursuit that followed, the Middlesex Artillery, being thinly manned, remained in the rear. Surprisingly, no casualties were recorded for Fleet's Battery.

On August 11, Jackson's forces were again on the move, retiring toward Orange Court House. On August 18, the army advanced again, crossing the river at Somerville Ford in pursuit of the Federal army, still under General John Pope.

By August 20, Jackson's men had reached the Rappahannock, marching through a driving rain to a campsite near Warrenton Springs. Living on short rations, the gunners of Walker's battalion remained near the same spot for the next few days.

The Federals appeared opposite the Confederates on the opposing bank of the river on August 24. With infantry lines being formed and batteries coming up to unlimber, the goal was recognized easily as the wooden bridge across the river. As the Federal batteries began to open, the Confederate guns were ordered to remain silent until the enemy infantry made a dash for the bridge. All of Walker's guns but Crenshaw's were posted on the heights near Warrenton Springs, where they had moved on the previous evening. McIntosh was on the right, followed next by Latham, and Fleet (still under Lieutenant Hardy). On the west side of the road leading from Jefferson to Warrenton Springs were the guns of Braxton and Davidson.

The Federals only made feeble attempts throughout the day, each being easily repulsed by the Confederate guns. That evening, the batteries began to withdraw from the heights and moved just behind the lines at Jeffersonton.

After the withdrawal to Jeffersonton, the trail of history for the Middlesex Artillery drops off abruptly. Whether they followed the army in its movements or not is uncertain. The numbers of the battery had fallen gravely below the standard for an artillery battery. By the opening of the Maryland Campaign the Middlesex Artillery had been moved to camp in Leesburg to begin recruiting. At least one battery member was detached in some manner during the advance to Sharpsburg. Subsequently, Private Silas Chandler was wounded and captured at Sharpsburg. Though wounded by a minie in the left knee, he would be released from capture and recover to later serve with the Ashland Artillery.

Though not active in the field, controversy began to take up a large amount of the unit's history following Warrenton Springs. General William N. Pendleton reported of the battery:

The battery is justly eulogized by its division commander and his chief of artillery for gallantry in several actions, and yet I submit respecting it these facts: It was sent to Leesburg, with a note from General Jackson, as not in efficient condition to move forward, and under the order there issued by direction of the commanding general for supplying four batteries, thus sent to the rear men and horses needed for the Fall efficiency of those sent forward. Eleven of its horses had been detached by the division chief of artillery. Lieutenant Hardy called on me to report and complain that he was thus detained, and when informed that he must submit, under the rules prescribed by the commanding general, bore himself with strangely improper violence, stalking out of the room where I was and slamming the door, and gate even, furiously. Passing over this, I the next day directed a detail of men and horses to be borrowed from the battery and assigned to two or three other batteries about to proceed into Maryland. The officer to whom it was committed returned and reported that Lieutenant Hardy refused compliance. Although extremely unwell, I immediately rode to the camp, and handing the order to Lieutenant Hardy, asked if he refused to obey. He promptly replied that he did refuse. It was then presented to each of the other lieutenants with the same result. They were immediately placed under arrest, and the refusal in the presence of the men being really mutinous, they were sent under guard to a distant part of the camp. An officer near was then pointed out as the captain. He informed me that he was under arrest by General Jackson. The orderly sergeant was then called for; the company had none, nor any sergeant. No roll of the company could be produced. I had the company formed. Some 50 men appeared nearly half of whom complained of being sick. I had them examined by my surgeon. He excused about four. The men then presented a number of requests and complaints. Many had been in service for seven months without a cent of pay, and bounty due had never been paid. No pay-roll or descriptive lists could be found. I examined the horses; they were in wretched condition, and, finally, 46 men, some volunteering for the purpose, were borrowed, and assigned to Captain Woolfolk, now with Col. S. D. Lee; to Capt. M. Johnson, now with General Gregg, and to Captain Ancell, now with Major Nelson. A few horses were also borrowed....

On receiving notice a day or two ago that General A. P. Hill thought highly of this battery and wished it restored,

I notified Major Richardson that the battery should prepare to move forward as soon as possible; that the lieutenants, if suitably atoning to violated discipline by a proper acknowledgment, should be released, and the battery coming forward would receive its borrowed men at once and occupy its old position; otherwise the delay of a trial must occur. They replied, defending their course. Their note was pronounced unsatisfactory, and a new opportunity given. They again, while disclaiming personal disrespect, maintained that their disobedience was a necessity. Their note was returned and a third opportunity given. Further answer they decline. The case, therefore, must await a court-martial. Charges shall immediately be submitted.

The commanding general will judge whether the principles first announced warrant the continuance of this battery, or require its being merged like those before mentioned. If the battery be thus distributed, I would recommend that its men, heretofore temporarily assigned, remain where they are, and that the remainder be assigned to Capt. M. Johnson, horses being turned over to the quartermaster and guns to the ordnance department. In all these cases, companies to which the men are assigned to have choice of the guns in the battery merged, and all public property in the latter not assigned to be turned over to the proper department. To insure the speedy and proper consummation of these measures, I recommend the appointment by the chief of artillery of two or more officers, who shall supervise the whole process.

The Middlesex Artillery, under strength and its officers arrested for insubordination, was consolidated with the Ashland Artillery and with Captain Valentine J. Clutter's Jackson Flying Artillery under the provision of Special Orders No. 209, Headquarters, Army of Northern Virginia, dated October 4, 1862. In all, 53 men were assigned to Clutter, 20 to Woolfolk, and 12 to Kirkpatrick. Though Hardy would have his court martial, he was assigned as enrolling officer for Essex County in December.

Hanover Light Artillery

Nelson's Battery 1861

Organized mostly of schoolboys from Hanover Academy, the Hanover Light Artillery was organized at Hanover Junction, now Doswell, in April 1861. Led initially by Captain William Nelson, the battery entered state service at Richmond on May 21, 1861. In the period between organization and mustering in, the greatest event for the battery's history occurred when young Henry Robinson Berkeley went to Hanover Junction to attend a rally and enlist in the ranks of the company. Had it not been for young Berkeley's diary, the history of the battery would have been left sorely wanting.

Within two days of young Berkeley's enlistment, he returned to Beaver Dam Depot, as many others did, to obtain uniforms. These uniforms were described in the company's muster rolls as being "good and substantial fatigue uniforms" provided at private expense. The following day, May 20, the company assembled and made its way to Richmond via train from Anderson's Crossing. Reaching the city late in the afternoon, the men went into quarters in Trinity Church on the corner of 15th and Franklin Streets near the Exchange Hotel. The officers saw to it from their own money that the men were fed breakfast and supper from the Columbia Hotel. The men were left only to find dinner for themselves.

On May 21 the battery was officially mustered for one year into state service by the Inspector General of Virginia troops, John B. Baldwin of Augusta County. The same day, the battery received its first orders -- to report to Jamestown Island.

On Friday, May 23, the company took the steamboat Curtis Peck down the James River. Arriving near noon, the company was ordered to new quarters in a large barn (probably that of Colonel Allen) east of College Creek and just below the island, having not yet received their tents. The men got Colonel Allen's slaves to fix them dinner. At their new station, the company began forming messes, Berkeley joining in the same one as his older brother, John Berkeley, and John Cooke, Frank Hall, George W. Nelson, Henry Stringfellow, Beverley Turner, and Dabney Williamson. The mess would grow the following month when Edmund Anderson, Robert Stringfellow, John Hall, and Leonidas Points joined.

For the duration of the time until the fall, very little occurred. Learning to cook in the field and drilling became the daily order of business. Only on June 10, during the excitement caused over the engagement at Big Bethel, and following the First Battle of Manassas, did the battery get disrupted in its peaceful living situation. Though many became concerned about missing the war while sitting inactive on Jamestown Island, Robinson

Berkeley and several others were quite content not to affix their names to a petition asking reassignment to Manassas. Throughout the tour on the island, the rolls reported the condition and discipline of the men, noting that their drill as only fair. Well into the fall, the company maintained no public property with the exception of a few camp kettles, axes, spades, and picks. According to Robinson Berkeley, the battery's guns during this time consisted of a "water battery" of four 64-pounder.

Soon the drilling became a point of grievance for the men of the Hanover Artillery. Captain William Nelson was strict and a stern disciplinarian, apparently seeing that the men were always given a full day's work. Over the duration of their time on Jamestown Island, the stern captain would become greatly resented for his role and would lose the re-election to the post the following spring. Though Captain Nelson received a petition from the company on October 1 calling for his resignation, he refused and continued his stand on discipline and order. In time, Nelson had a great many other achievements waiting for him beyond the Hanover Artillery, being a man who was also sincere and kind, but more importantly noticed by many senior officers for his qualities as a potential field officer.

Within two weeks of the petition, Robinson Berkeley developed a special appreciation for Nelson when the captain personally took charge of him on his trip to Richmond. Sick with typhoid, Berkeley ran a high fever and was unable to move on his own. Nelson carried the young man from their difficult boat journey to the Powhatan House Hotel in Richmond. From the hotel, the trip continued on October 15 when Berkeley was put on a train under the care of Dr. Prosser Harrison until reaching his home in Hanover County.

Though Berkeley recovered, his time apart from the battery leaves a gap in the history of the battery's first and last winter quarters. Berkeley did record, after his return, that the battery had moved in November to Yorktown and been given six field guns with horses and the necessary equipment. From there, the Hanover Artillery had gone on an expedition to Mathews Court House, only to return a short time afterwards and go into winter quarters a mile below Yorktown. It can be assumed that the Hanover Artillery spent the duration of the winter without event.

The Trial By Fire 1862

On January 31, the Hanover Artillery was recorded as in the vicinity of Ship Point. There the battery was placed in a brigade under Brigadier General G. J. Rains along with the 13th Alabama, 2nd Florida, 6th and 23rd Georgia, Louisiana Zouave Battalion, 2nd Mississippi, 15th North Carolina, 32nd and 53rd Virginia, 115th Virginia Militia, and Maurin's Louisiana Battery. At Yorktown, sickness continued to take its toll in the ranks. Samuel Luck died on February 11.

On March 8, young Robinson Berkeley had recovered enough to rejoin his company. He made his way into Richmond, had two photographs taken, and re-enlisted. On March 13, Berkeley and a few other furloughed members returning to their company arrived a mile below Yorktown to find that the company had moved toward Bethel on a scouting expedition. They remained alone in the camp for six days; the battery returned on March 19.

The Hanover Artillery was ordered into the works at Yorktown on March 28 in preparation for a spring offensive by the Federal army under General George B. McClellan. Remaining until April 2, the battery was moved again out of the works at Yorktown and ordered to Dam Number 1 on the Warwick River There the battery remained for but a few days, defending the potential crossing from the enemy. On the evening of April 4, Nelson's Battery moved into the "White Redoubt" south-southwest of the Yorktown defenses.

In their new position, the Hanover men drew their first taste of enemy fire when the Federals opened their guns in the morning. While the bulk of the battery remained at the White Redoubt under fire for nearly two hours, one gun, under Lieutenant George W. Nelson, remained stationed at Dam Number 1. After dark, the entire company was again ordered back to Dam Number 1.

The following day at 11 a.m., the Federals again advanced on the Hanover Artillery's position. Berkeley at a party of Federal cavalry and watching the enemy horsemen fall under the blast of the 24-pounder shell. One Yankee cavalryman did come within 200 yards of the guns and captured the Confederate picket, though not from Nelson's battery. On April 7, the enemy again advanced and was identified as Vermonters by Berkeley. After the Hanover men fired several charges from the cannon, the enemy troops were dislodged from their position in a thicket a few hundred yards from the guns.

For the next four days, the Hanover Artillery remained on line near Dam Number 1, returning fire as necessary to thwart any advances by enemy troops. On April 12, the battery again returned to the lines near Yorktown, placing the guns at various points along the works. Nearly every day for the next few weeks, the battery came under fire from Federal gunboats. While no men were killed or wounded, Private Shelbourne received a fright when a large shell burrowed in under his tent several feet into the ground and blew him into the air. Shelbourne quickly recovered, having been saved by the fact that the shell had exploded so deep in the ground.

On April 15, the rifled piece of Nelson's Battery was ordered to Gloucester Point. It apparently returned soon after without event. The following day, some of the battery's guns were apparently sent back to Dam Number 1, where an assault was received yet again. With only one 24-pounder, Captain Nelson occupied the front and most exposed position immediately at the dam. Meanwhile, two pieces of the Troup Artillery under Captain Stanley occupied positions at the right and left redoubts, about 200 yards to the rear on rising ground. The few Confederate guns were met with a terrific fire from three Federal batteries that opened on them with "a converging fire of sixteen guns." The enemy again attempted an assault across the dam, which was eventually repulsed by Confederate infantry with the aid of the artillery.

While the lines around Yorktown lit up with firepower daily, the Hanover Artillery still had time to reorganize. On April 30, under the provisions of the Confederate conscription act, the battery held an election, removing Captain William Nelson from his post and raising George Washington Nelson. William Nelson was soon after promoted to major and led a battalion of artillery. W. M. Fontaine, Henry M. Stringfellow, and William Basye were elected to lieutenants. Nearly all of the non-commissioned staff changed as well.

On May 2, as McClellan continued to apply pressure to the Confederate lines along the peninsula, the Hanover Artillery evacuated with its brigade after nightfall. Ordered to follow the 8th Georgia, the battery had hardly moved out along the Williamsburg Road before being recalled into position due to the congestion along the road. The following day, the battery again attempted to withdraw, again pulling out at dark. After following the brigade along a muddy and dark road, the battery reached the west side of Williamsburg sometime in the night.

Two days later, the battery again continued toward Richmond. Under a heavy downpour, the Hanover Artillery trudged along until called upon to send two of its lightest guns with the brigade to the battle at Williamsburg. Near sunset, all of the Hanover Artillery was ordered to the battlefield. No guns of the battery, however, saw any action during the day.

For the next few days, the Hanover men continued, under miserable conditions and with little food, toward Richmond. On May 18 the battery reached the New Market Road, one mile below the city, and went into camp. Within a week the battery was again moved, to a Mr. Davis' farm. Remaining there for almost another week, the battery was not called upon to serve until May 31, when ordered to accompany the brigade to the action at Seven Pines. Due to the weight of the heavy guns, however, the company found it impossible to move through the deep mud from the junction of the Williamsburg and Charles City roads. The Hanover Artillery therefore, remained at the junction until the conclusion of the battle.

On June 2, Nelson's Battery went into camp near the junction on the John Poe farm. The next day, the Hanover Artillery received lighter guns in exchange for the 24-pounders and was ordered to rejoin the brigade three miles down the Williamsburg Road. Stopping overnight along the way, the battery did without blankets, having regrettably left them at their former campsite. The following day the company went into yet another camp until June 24.

The company moved on Wednesday, June 25, from its campsite. The men reassembled on the Williamsburg Road and moved to a new position where they remained in battery until dusk. Again the Hanover Artillery saw no action and was returned to camp, only to be moved back into line again with the brigade near midnight. Near Mechanicsville, the battery awaited orders to move into battle, but was again disappointed.

Back in camp again on June 28, Nelson's Battery took to foraging for rations. The following day, the company resumed the march toward Mechanicsville and Cold Harbor, passing swollen and unburied corpses from the recent battles. The bridge repaired across the Chickahominy allowed passage with the brigade on June 30. The company toured the battlefield and gathered assorted necessary blankets, haversacks, knapsacks, oilcloths, and tents. Most men did not weigh themselves down with much booty for fear that they would soon go into action. Indeed the battery was called upon the same day.

After being allowed time to gather some of the much needed supplies, the battery resumed the march. Sometime in the afternoon, two of the Hanover Artillery's guns were called up with skirmishers from the 6th Georgia under Colonel A. H. Colquitt. Advancing slowly, the contingent gathered several prisoners along a route littered with muskets, bayonets, and assorted equipage. Finally reaching White Oak Swamp, the advance came to a halt upon finding the Federals in line of battle on the opposite side of the swamp.

After being received by fire from the enemy cannon and sharpshooters, artillery officers began an examination of the area to find sites employment of the Confederate batteries. A plan was soon devised and a rough road was cut into the woods to the right of the road. Seven batteries, including the Hanover Artillery, were positioned unseen by the enemy, "behind the crest of the hill on this side, about 1,000 yards from the enemy's batteries and some 1,200 yards from their infantry." At 1:45 p.m. the batteries opened and surprised the Federals. After returning the fire briefly, the Federals were believed to retire from their position. However, shortly thereafter, approximately eighteen Federal guns opened and began a duel that lasted until dusk. In the action of the day, four horses were killed from the Hanover Artillery.

The battery retired from the field at about 4 p.m. having used up all of its ammunition but a few rounds of canister. Other batteries came in to take up Nelson's previous position. Along the way to the rear the Hanover men watched as "Stonewall" Jackson's men marched to the front.

Bivouacking for the night, the battery was again ordered to move on July 1 to Fair Oaks. Reaching there by noon, the Hanover Artillery went into a former Federal campsite on the right of the road to Richmond; all of the artillery from General Hill's division having been ordered to rest near Seven Pines that day. While the tired cannoneers rested they could hear the guns of their brother batteries go into action at Malvern Hill. The following day the battery remained in camp while the limbers and caissons were sent to Richmond for replenishment. Not until July 3 did the Hanover Artillery rejoin the brigade, which was found along the Charles City Road. Camp was then made on the Brackett farm on the north side of the Charles City Road, nine miles from the Williamsburg Road.

On July 8, Nelson's Battery accompanied the brigade and Hampton's cavalry on an expedition to the Gatewood farm near Malvern Hill. Remaining for a day, the contingent then returned to the campsites near Richmond. The Hanover men would receive a long rest for over a month before serving actively again. Life in camp was only disturbed by the occasional rumbling of the first section's guns and caissons into their picket post on the White Oak Swamp. The swamp air apparently began to take a toll while in camp. Berkeley recalled that several horses died from ill care, as well as Private Claudy Anderson from typhoid.

As had been the case with Captain William Nelson, Captain George "Wash" Nelson became the butt of company jokes and abuse. While not so stern a disciplinarian as the former, "Wash" was proving himself incompetent enough to convince several members to ask for transfer to the Hanover Cavalry. Naturally, the captain turned the majority of the men down in their requests. Only Jack Cooke was permitted the transfer, while Frank Hall was granted a transfer to the Fredericksburg Artillery.

Not until August 15 was the battery again ordered to move with the brigade, now under Colonel A. H. Colquitt. The horses were in such poor condition that even the march to Hanover Junction proved nearly impossible, leaving Captain R. C. M. Page's Morris Artillery to take their place. Meanwhile, the Hanover Artillery again came under the command of William Nelson, now major, in his artillery battalion. On August 18, the battery moved to Major Nelson's camp near Oakwood Cemetery and reported for duty. On the following day, Captain Nelson appealed to the company for tolerance, admitting that he had not done his full duty as commanding officer and hoping to do better. Doubt still ran rampant through the ranks. Moving to Hanover Junction on August 20, the Hanover Artillery spent an enjoyable week in the vicinity of their homes before being ordered out again.

By 10 a.m. on August 26, the battery began its march with the army, reaching Louisa Court House on August 28, near Rapidan Station on August 29, Culpeper by August 30, and across the Rappahannock River and to Warrenton Springs by September 1. Upon reaching Warrenton Springs, the Hanover Artillery horses had totally given out, forcing the battery to remain behind with orders to rest the horses for one day before following the command to Leesburg. On September 3, the march was resumed, encamping that night at Gainesville on the Manassas Gap Railroad before heading out the next day to Aldie, near Mt. Zion Church, and finally reaching Leesburg on September 5 and going into camp on the Winchester Pike. Along the way the rations had been extremely short, leaving the men to scavenge whatever was possible en route.

On September 7, the sign of things to come began when all of the best horses from the battery were taken with other batteries across the Potomac. The following days were spent carrying guns, wagons, and other plunder to Winchester with the remaining broken down horses. On the night of September 8, Nelson's company moved with the battalion toward Winchester under a brilliant moon, encamping after twelve miles on a farm near Purcellville.

While in camp for the next few days, the men subsisted on wonderful food provided at extremely low prices by the folks of the surrounding countryside. Robinson Berkeley recalled that it was the best he had ever had the opportunity to eat and made him gain a few pounds. On September 11, the march continued again, crossing the Blue Ridge at Snicker's Gap before encamping on the west side of the Shenandoah River. The men of the extra companies, including the Hanover Artillery, were utilized to move the extra guns with the ever increasingly fatigued horses. Continuing the march, the company camped at Spout Spring on September 13, and finally reached Winchester on September 15, where the extra guns were turned over to Major John A. Harman, the quartermaster for Jackson's command.

The laborious job completed, the men settled into the beautiful fal! of the Shenandoah Valley, camping at Spout Springs on the Opequon again for a week before moving out on the Valley Pike to a camp just beyond Kernstown, on the east side of the pike.

Though disbanded by Special Orders Number 209 on October 4, the Hanover Artillery did not hear of it until the following day. Most of the 60 men of the Hanover Artillery were reassigned to the Amherst and Ashland Artillery companies. Captain Nelson was thought better suited to be a cavalry aide than battery commander. "Captain Nelson has served with as distinguished gallantry as any officer in the Confederate Army. For cool intrepidity and heroic daring - indeed, for the most genuine fidelity and chivalry - Capt - Wash. Nelson is unsurpassed. He has earned high eulogiums on the battle-field, and yet he is not, in some respects, adapted to take care of a battery, nor are his lieutenants, though one of them (Lieutenant Fontaine) is a good officer to be provided for." Thus concluded the roll of the Hanover Artillery.

Both William and George W. Nelson went on to serve as officers in Virginia throughout the war, but one officer was confronted in the last days of the battery for his actions with the battery. In a petition initiated by the men of the Hanover Artillery prior to the disbanding, Lieutenant Henry Martyn Stringfellow was asked to resign. It was presented to him on October 6, the day following the news of the disbanding. John Hill and Sam Mills both confronted Stringfellow with the petition. In the ensuing argument, Stringfellow was accused of skulking from all of the battles around Richmond except White Oak Swamp. Even then, Stringfellow, when called to the battle, was found hiding in a gully in the rear of his guns. After the confrontation calling him a liar and coward, Stringfellow could not and did not remain in Virginia after the disbanding of the battery. He went to serve in Texas under General John B. Magruder as an ordnance officer. There, Stringfellow gained only one appraisal in the Official Records, while in the Galveston Campaign. As a result of his heroic actions in taking charge of some guns after the officers of a battery had all been wounded, Stringfellow was recommended for promotion to captain by Magruder.

Henry Robinson Berkeley continued his diary throughout the war until 1865. Serving in Kirkpatrick's company, Berkeley was captured in the Shenandoah Valley in 1864 and spent the rest of the war in the prisoner-of-war camp at Fort Delaware. Berkeley resumed civilian life after the war, marrying the girl to whom he had given his Literary Badge days after having enlisted in the Hanover Artillery. In 1886 the Berkeleys' relocated to

Orange County, where Robinson taught his own school. He died as one of the historians of Lee's Long Arm in January 1916.

The Magruder Light Artillery

Page's Battery 1862

Organized March 31, 1862, the Magruder Light Artillery was named for General John B. Magruder, then defending the Virginia peninsula. Enlisting for two years, the men who had joined from various other organizations had no idea that their battery would not even last seven months.

The first mention of the battery in any records was made in April 1862. With one section stationed at Dam Number 1, the men of the Magruder Artillery stood with Captain Thomas L. Rosser's Washington Artillery, Richardson's Battery, and a section of Palmer's howitzers. Apparently the rest of Page's Battery was then stationed at Redoubts Number 4 and 5, in the curtain connecting these redoubts, at Yorktown, or somewhere in the intermediate positions.

Throughout the Peninsula Campaign, Page's Battery remained in General Magruder's division. Under Colonel S.D. Lee's battalion, the Magruder Light Artillery served with Kirkpatrick's Amherst, Read's Pulaski (Ga.), and Richardson's batteries.

On the evening of June 26, Page's Battery was moved into position from where it opened on enemy works near the Garnett house. With the batteries of Lane and Woolfolk, the guns broke up the Federal working parties, but were themselves forced to withdraw by enemy batteries. Though the battery was not listed in action on the following day, Private Frederick Schneider was listed as killed and 1st Sergeant Robert Bledsoe was recorded in the Combined Service Record as wounded.

By June 30, the Magruder Light Artillery was at Holly Springs. Since the departure of General Magruder, the battery had grown very sparse in numbers. In July, General Magruder requested that Page's Battery be sent to join him in the Trans-Mississippi Department. In his letter he wrote:

> I have the honor to request that Capt. T.J. Page's battery be ordered to report to me. This battery is composed of three small light rifled pieces and was organized by me in the Peninsula. The horses, which are very fine, can be turned over to the quartermaster here if transportation cannot be afforded them. This battery is not needed here, there being more artillery than the army requires. I desire that three more guns may be added to it to make it an efficient battery. It is now but a fragment of a battery.

Magruder's request would not be fulfilled and the Magruder Artillery remained in Virginia. Little else is known of the battery's movements through September. On July 25, Page recorded in the muster rolls that the battery was camped near Nine Mile Road.

Between 9 and 10 a.m. on September 14, the Magruder Artillery was finally again put in action. With the enemy advancing toward Crampton's Gap in Maryland, the Confederate forces there needed artillery support. A report written by Colonel E. B. Montague of the 32nd Virginia Infantry described the event:

> At about 11 o'clock they masked the most of their force under a hill and wood about 3 miles, and advanced two brigades by the left flank into a field opposite our position. Meantime I had sent to General Semmes for re-enforcements, and he promptly ordered up the Fifty-Third Georgia Regiment and three pieces of artillery (rifled), under the command of Captain Macon, two of his own guns, and one of Captain [Lieutenant G.A.] Magruder's.

By 4 p.m., the Federal forces withdrew their skirmishers and began a move against the Confederate right flank, "leaving Burkittsville on his left..." By moving the artillery to the left a few hundred yards, the enemy could be brought under range of the guns. Subsequently, Macon's, Manly's, and Magruder's guns were relocated and "played upon the enemy with great effect, time and again their ranks being broken by their deliberate and well-directed fire, the enemy's guns not being able to reach us on account of our elevated position...."

Though the service of the battery had been effective at Crampton's Gap, it was soon scheduled for disbandment along with several other artillery companies of the Army of Northen Virginia. General Pendleton wrote on the last chapter of the history of the Magruder Light Artillery:

> Capt. Thomas Jefferson Page's (jr.) battery, as below the service standard, with little prospect of increased strength, having only about 45 men and 25 horses. Captain Page is a brave and faithful officer, and his company has in several actions done good service. He would make an excellent staff officer, and as such I cordially recommend him. Lieutenant Magruder also deserves well, and will, I hope, have a suitable position. The men and horses I recommend for assignment to the Virginia companies under Col. S.D. Lee, and Captain Page himself to be attached to that command in the capacity of aide to the colonel.

The Manchester Artillery

The "Bridesmaid" Battery 1861 - 1862

The Manchester Artillery was organized at Manchester, Chesterfield County on May 1, 1861, by Captain William R. Weisiger for one year. Assigned as Company I, 16th Virginia Infantry, it assumed positions at Sewell's Point on the Chesapeake Bay by mid-May 1861. Sometime soon after it was reorganized and issued 100 flintlock muskets. This company did not become an independent battery until April 30, 1862.

Like many of the batteries that were disbanded in 1862, the Manchester Artillery has a limited history of service. Though the battery does not show an unofficial merger with the Courtney Artillery until September 1862, Weisiger's Battery apparently followed the Courtney Artillery as a sub-organization throughout 1862. There is no documentation to show this officially other than proof illustrated in the Compiled Service Records. For instance, the Manchester Artillery, while not listed as an active unit in the 1862 Shenandoah Valley Campaign, had casualties at the battle of Cross Keys on June 8. Thomas J. Cox and John Shell were killed in action while Cornelius Hubbard was listed with a wound in the arm. The following day another man was wounded at the battle of Port Republic: James A. Vaden died twenty-one days later of his wounds at the University of Virginia hospital in Charlottesville.

For unknown reasons, the company's organization seemed to melt apart soon after the battles of Cross Keys and Port Republic. On July 4, both 2nd Lieutenants Nathan Eubanks and Benjamin Weisiger were relieved. Then on July 15, Captain Emmett Weisiger resigned. The muster rolls offer little more until July 24, when the battery was recorded in camp near Gordonsville, Virginia. Organized under General Richard S. Ewell's division on July 31, the battery fell under Brigadier General Isaac R. Trimble's Brigade, along with the infantry regiments of the 15th Alabama, 21st Georgia, 21st North Carolina, and 1st North Carolina Battalion Sharpshooters. In addition to these, the battery would serve alongside Captain Joseph W. Latimer's Courtney Artillery. Ultimately, the battery history of the Manchester Artillery should probably parallel that of the Courtney Artillery up until its semi-official merger with that company in September. Scant data is recorded of the history of Weisiger's company.

On August 19, William Robert Weisiger succeeded Emmett Weisiger in the captaincy of the company, Emmett's resignation finally having been approved. The title of captain may have only been on paper, as William was shown as a surgeon, separate from the Manchester organization at about the same time. The official merger with the Courtney Artillery did not come until 1863. However, the battery was merged several times on paper. The first unofficial merger appears in August, followed by yet another, on paper, on September 11. The company was obviously being dissolved off the record as the last senior lieutenant, 1st Lieutenant Thomas Gibbs was relieved on September 1.

Carrying its own muster sheets, the Manchester Artillery continued to record movements until officially merged in 1863. On October 31 the company was recorded near Ripley, Clarke County, Virginia; on November 27 it was near Madison Court House; and on January 1 and February 28 it was near Port Royal, Virginia. The last casualty listed in the Manchester Artillery rolls was Private Atlas Pond, who was mortally wounded on December 13 at Fredericksburg. Lingering for over a month, Pond died on January 22, 1863. Finally, by Special Orders Number 92, Adjutant and Inspector General's Office, dated April 15, 1863, the company was officially disbanded and men, guns, equipment, and horses turned over to the Courtney Artillery. While always in support of the Courtney Artillery since its organization as an independent battery, it had always been a bridesmaid to the Courtney unit, never gaining full separate recognition for its service in the field.

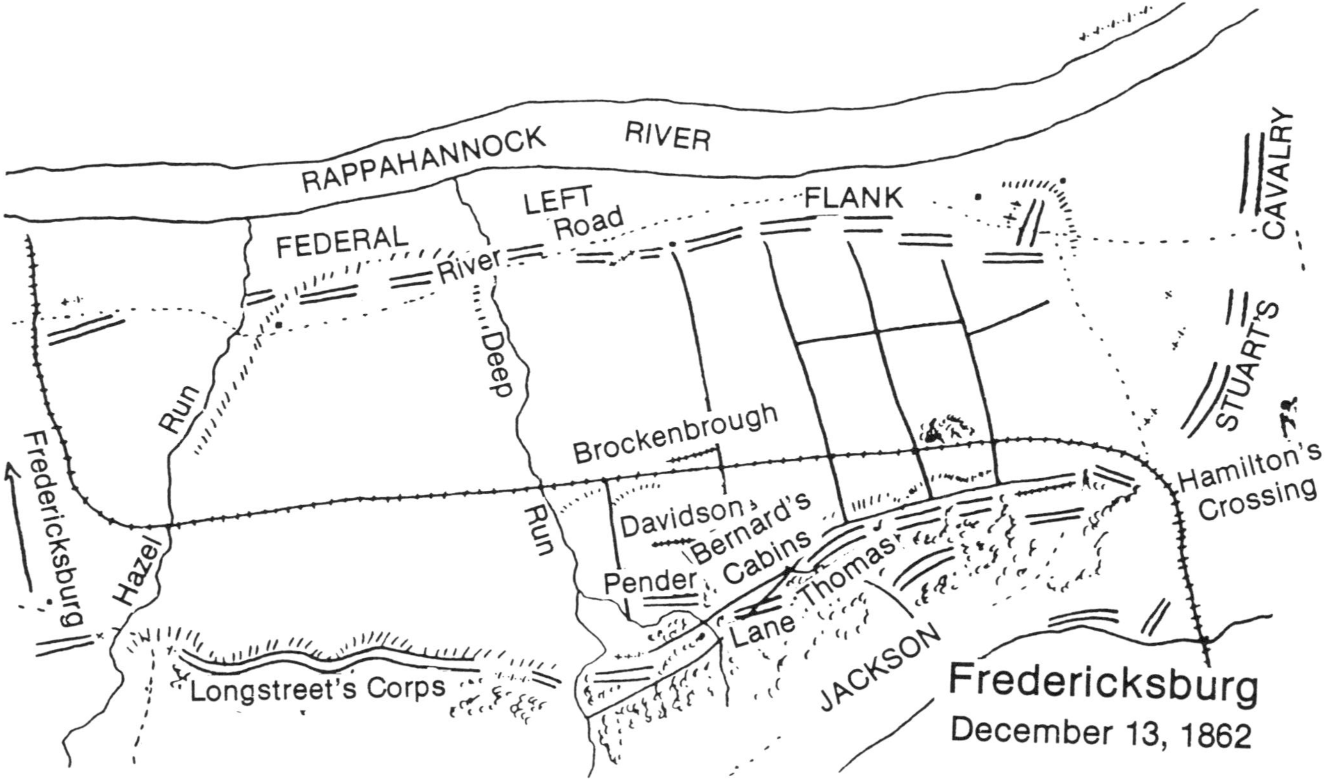

RAPPAHANNOCK
RIVER
FEDERAL
LEFT
FLANK
River
Road
CAVALRY
STUART'S
Run
Deep
Brockenbrough
Hamilton's
Crossing
Fredericksburg
Hazel
Run
Davidson
Bernard's
Cabins
Pender
Thomas
Lane
Longstreet's Corps
JACKSON
Fredericksburg
December 13, 1862

The Campbell Long Island Artillery

A Year Against Disease and Enemy Fire 1862

Patrick H. Clark's Company, also known as Abram Wimbish's Company or the Campbell Long Island Artillery, was organized on March 15, 1862, in Campbell County. The company mustered into service on March 20, 1862, and was intended to be Company E of Fitzgerald's 1st Confederate Artillery. This regiment was never fully organized and the Long Island Artillery became an independent battery. The Long Island Artillery subsequently became part of Major Hilary P. Jones' Battalion of Reserve Artillery, Army of Northern Virginia.

The battery spent its first two months in and around Richmond training and preparing for battle. Disease took several lives in the company during this period. On May 5, Private T. G. Woodall died, followed in two days by his brother J. H. Woodall. Within a week later, the company lost Asa Hamlette, his brother Lee Hamlette and Cabell Holt. May's last recorded death was on the 15th when Thomas Grisham passed away. It would be another two months before death would again settle on the camp of the Long Island Artillery.

Though down to only 48 men, the battery remained attached to Brigadier General Howell Cobb's Brigade, serving with the 16th and 24th Georgia Infantry, Cobb's Georgia Legion, the 2nd Louisiana, the 17th Mississippi, and 15th North Carolina Infantry.

By May 23, Clark's four-gun battery with the batteries of Rhett, Stamp, Peyton, Moseley, Forrest, and Lloyd, was ordered to report without delay to General Joseph E. Johnston for field service. Though required for immediate service, the Long Island Artillery was still below standard in numbers of men and equipment, leaving the company out of action and rarely mentioned in reports. However, Private William Johnston was recorded as being wounded at the battle of Seven Pines.

On June 17, General Pendleton's headquarters issued General Orders Number 1, which assigned Rhett's, Clark's, and Peyton's batteries to a field battalion commanded by Major Hilary P. Jones. Jones' Battalion was to act in reserve with General Daniel Harvey Hill's division. When these orders were issued, all three batteries were stationed at or near Redoubts 2 and 3 in the Richmond defensive perimeter. While in this organizational arrangement, Clark's Battery fought with General D. H. Hill's division on Dr. Garnett's farm on June 25. Muster rolls indicate that James A. St. John was wounded in the side. Major Jones' report on the engagement gave a detailed account of the activities of the batteries under his command:

Having been assigned with the batteries of Capt. A. B. Rhett, P. H. Clark, and Jefferson Peyton (that of the latter under command of Lieut. C. W. Fry, in the absence of the captain, who is sick) to act as reserve to General D. H. Hill's division, we left our camp, on the Williamsburg road, on the night of Wednesday, 25th ultimo, with the other batteries of the division, marching in the direction of Mechanicsville.

Captain Rhett had previously been ordered to report for duty with General Ripley's brigade, and with it, about 4 p.m. of Thursday, crossed the Chickahominy in advance of other troops of the division. He experienced some difficulty in crossing on account of the destruction of the bridges over the stream by the enemy. He succeeded, with the help of the Pioneer Corps, in rendering the bridge passable, and crossed with his batteries and engaged in a very spirited manner the enemy's batteries, which he continued to do, changing his position whenever he found that the enemy had his range, until ordered to cease firing, about 9 p.m. In this engagement he suffered from a cross-fire of the enemy, and had 11 men wounded and 6 horses. As soon as the passage was opened I crossed with the two remaining batteries and bivouacked for the night near Mechanicsville.

The next morning we were awakened by a few shots of the enemy, which passed over us without doing any dam an order to send a battery to the top of the hill in rear and protected by one of the enemy's redoubts, I ordered Captain Clark to this position. Here, with other batteries of the division, he fired several rounds at the earthworks of the enemy on the opposite side of a ravine in front of us, but received no reply.

The division then took up the line of march with the reserve batteries in the rear. The line was halted at Cold Harbor, and Captain Rhett's battery was ordered to the front to support Captain J. W. Bondurant's battery, which was actively engaged with a battery of the enemy. I also ordered Captain Clark and Lieutenant Fry to bring their batteries up in easy supporting distance, sheltering them as much as possible from the fire of the enemy, which enfiladed the road, by placing them in a ravine to the left of the road. Subsequently I ordered the two reserve batteries on the right of the line, Captain Rhett retaining his position, but by a new disposition of the forces being on the left. This new disposition of the line enabled the batteries to open on the enemy, which they could not do before, owing to our troops being between the enemy's and our batteries.

Captain Clark and Lieutenant Fry were still held in reserve to support the attack of the infantry, and here they were exposed to an annoying fire of the enemy's battery, which was to our left and front. Captain Rhett's, with other batteries of the division, engaged the enemy's battery and soon succeeded in silencing it. In this he had 2 men and 1 horse killed and 3 wounded. We again bivouacked for the night in the position we had fought.

At 3 a.m. of Saturday, we were ordered to move forward to a position at Dr. Gaines' house, where we remained, expecting the enemy's batteries to open on us, until 12 m., when we went forward and took position on the hill overlooking the approach to Grapevine Bridge. Here we remained until 2 a.m. Monday, when we commenced to cross the stream, the bridge over which had been destroyed by the enemy and had to be reconstructed by our men.

We continued the pursuit of the enemy until about 12 or 1 o'clock in the day, when we came up with him at White Oak Swamp. Here he occupied a position on the hill opposite, with twelve pieces in sight. All the batteries of the division and of the reserve were ordered to the front and engaged the enemy, forcing him to change his position in a very short time, which he did, leaving three of his guns disabled on the field. Taking up a position to the right of his former one, or having other batteries there, he again opened on us, and the firing continued obstinately, with slight intermissions, until night. In this position the batteries of my battalion were more exposed on account of the nature of the ground, and consequently suffered more than others. I should mention that Colonel Crutchfield was in command on this occasion.

Captain Rhett lost 2 men killed and 3 wounded; also 1 horse killed and 2 wounded. In the three engagements he expended 1,320 rounds. Captain Clark lost 1 man killed and 5 wounded [all at White Oak Swamp]. Two wheels were seriously damaged. Captain Clark at Mechanicsville and White Oak Swamp expended 392 rounds.

Lieutenant Fry had 2 men wounded and 3 horses killed and fired 390 rounds of ammunition....

It will be seen from this report that Captain Rhett's battery was in three engagements, and the other two, although in but one, were still exposed to the fire of the enemy on another occasion.

I cannot speak in too high terms of the gallantry of the officers and men of the battalion--not only their gallantry on the field, but the cheerful spirit with which they endured the hardships of the eventful week. As a proof of the heroic and dutiful spirit of the men I cite with pride the fact that during the whole time there was not a single straggler from the ranks.

By July 3, the battalion had dwindled significantly in numbers. Clark's and Peyton's batteries together could only form one battery of six guns. Disease, specifically typhoid, again took a toll on the numbers in the Long Island Artillery. John Storer died in Richmond on July 5, while Z. M. Hyatt died three days later. On July 26, the battery's own Captain Clark died, followed on July 29 by the death of Robert Holt from fever.

After Abram Wimbish's election to the recently vacated post of captain, the battery's history again goes silent until July 31, when the Campbell Long Island Artillery took part in an action on Coggins Point, near Petersburg. Few details about the participation of the battery in the action are available. A number of Confederate batteries, including Clark's, engaged Federal shipping in the estuaries approaching Richmond.

Though it accompanied D. H. Hill's division to Sharpsburg, the battery remained out of action. Finally, on October 2, Brigadier General William N. Pendleton made suggestions for numerous changes in the Army of Northern Virginia's artillery corps. His comments on Major Jones' Battalion spelled doom for the Campbell Long Island Artillery:

The four companies of Major [H. P.] Jones' battalion be consolidated into two, thus: By relieving the officers of Captain [Abram] Wimbish's battery as not efficient, however otherwise worthy, and assign the men to Major Jones for distribution among his two companies retained....

About 28 members of this company were transferred to Captain Richard Channing Moore Page's Morris Artillery, while 61 men were transferred to Captain Charles William Fry's Orange Artillery.

Appendix

Tabulations From the Rosters

Note: Data provided herein is only the data available from the rosters. The reader will notice that some categories covered by one battery are not included in another. The reason is either that the material was not provided in the roster or there were none (e.g., MIA's, Wounded, Conscripts, etc.) recorded for that unit. Enlistment ages were available in only a few battery's records. The age and occupation listings will not be available for all batteries covered in this volume.

Commissioned and Noncommissioned Officers

Company A, 1st Regiment Virginia Artillery
Washington/Hampton Artillery

Fraser, William W., Captain
Smith, Charles L., Capt.
Semple, George William, 1st Lt.
Courtney, Alfred Ranson, Lt.
Daugherty, William T., 2nd Lt.
Hawes, Samuel Horace, 2nd Lt.
Hope, George Washington, 2nd Lt.
Hope, Jesse P., 2nd Lt.
Tabb, Thomas, 2nd Lt.
Guy, Adolphus, Sgt. Maj.
Adler, Abraham - QM Sgt.
Kendall, John W., 1st Sgt.
Lawson, John W., Orderly/1st Sgt.
Peek, John L., 2nd Sgt.
Peek, Edward K., 3rd Sgt.
Wood, Edward C., 3rd Sgt.
Jones, Isaac L., 4th Sgt.
Cake, Richard B.: 5th Sgt.
Jones, William S., 6th Sgt.
Davis, James, 1st Cpl.
Westheimer, Harris, 2nd Cpl.
Hubbard, Lepron, 6th Cpl.

Company C (1st Organization), 1st Regiment Virginia Artillery
Beauregard Rifles

Schaeffer, Frank B., Captain
Cummins, Edward H., 1st Lt.
Cox, George Thomas, 2nd Lt.

Hill, Charles S., 2nd Lt.
Anderson, Arthur H.: 3rd Sgt.
Dodson, Richard T., 3rd Sgt.
Bayley, William H.C.: 4th Sgt.
Long, Melchoir M., 5th Sgt.
Boone, John F.: 1st Cpl.
Smoot, Joseph K., Cpl.
Magruder, George, 3rd Cpl.

Company C (2nd Organization), 1st Regiment Virginia Artillery
Henrico Artillery

Ritter, William Bailey, Captain
Sands, Johnson H., Capt.
Robertson, Walter H., 1st Lt.
Crump, Sylvanus G., Jr. 1st Lt.
Crump, William B., 2nd Lt.
Burrows, Howard L.: Ord. Sgt.
Heilman, Julius F., Jr. Ord. Sgt.
Johnston, Thomas J., Ord. Sgt.
Smyth, Henry B., Sgt.
Lewis, Robert Eston, 5th Sgt.
Pollard, George B., 4th Sgt.
Flippin, Martin V., 1st Cpl.
Garnes, Albert, Cpl.
Graham, James, Cpl.
Kelly, George Washington, Cpl.
Regnault, Charles H., Cpl.
Shackleford, Lawson, Cpl.
Quarles, George Washington, 4th Cpl.

Company F, 1st Regiment Virginia Artillery
Lee/Williamsburg Artillery

Garrett, William Robertson, Captain
Coke, John Archer, Capt.
Hill, Charles, 1st Lt.
Lawson, James S., 1st Lt.
Pendleton, Samuel H., 1st Lt.
Robertson, William H., 1st Lt.
Richardson, Nathan, 2nd Lt.
Shield, Orlando F., 2nd Lt.
Palmer, Henry E., Lt.
Vanname, Peter M., QM Sgt.
Wilson, Josiah N., 1st Sgt.
Moore, Richard A., 2nd Sgt.

Potts, John, 3rd Sgt.
Wood, William C., 3rd Sgt.
Hazelgrove, John S., 4th Sgt.
Lipscomb, Spotswood, 4th Sgt.
Dodd, William Fletcher, 5th Sgt.
Jones, Isaac L., 1st Cpl.
Lindsay, William, 1st Cpl.
Roach, Christopher, 1st Cpl.
Caruthers, Robert: 3rd Cpl.
Charles, William H: 3rd Cpl.
Johnson, Cornelius, 4th Cpl.
Vandeveter, James, 4th Cpl.
Lindsay, Thomas H., 6th Cpl.

Company G, 1st Regiment Virginia Artillery
Peninsula Artillery

Cosnahan, Joseph B., Captain
Jones, William B, Capt.
Bright, Robert A.: 1st Lt.
Morrison, John H., 1st Sgt.
Higginson, W. Morehead, 2nd Sgt.
Moreland, Higginson W., 2nd Sgt.
Taylor, Nathaniel, Jr., 3rd Sgt.
Brown, Dixon, 1st Cpl.
Hogge, Addison, 2nd Cpl.
Chapman, John F.: 3rd Cpl.
Jones, Henry F., 4th Sgt.
Sherwell, William, 4th Cpl.
Pettitt, William F., 5th Sgt.
Presson, William M., Cpl.

Wise Artillery

Alburtis, Ephraim Gaither, Captain
Brown, James S., Capt.
Graves, Richard Perrin, 1st Lt.
Pelham, John, 1st Lt.
Walker, Elias M., 1st Lt
Kerns, Cyrus McLean, 2nd Lt.
Maxwell, John, Brevet 2nd Lt.
Murphy, George H., 2nd Lt.
Graves, William C., 3rd Lt.
Norman, G.S., Surgeon
Means, T.A., Asst Surgeon
Stewart, Bernard "Barney" W., 1st Sgt

Wentz, Henry D., 1st Sgt
Lucas, Benjamin, 2nd Sgt
Weddel, John H., Jr., 2nd Sgt
Tate, Robert J., 3rd Sgt.
Beard, George W., 4th Sgt
Beall, Richard H., 5th Sgt
Robinson, John S., Sgt
Wirt, Henry D., Sgt
Couchman, Jacob R., QM Sgt
Lantz, Joseph T. Van L., 1st Cpl
Wann, John A., 2nd Cpl
Lucas, Charles M., 4th Cpl
Seibert, Joseph, Cpl
Whitehurst, James, Cpl.
Langley, J. Lyle, Bugler
Sherrer, Joseph E., Bugler
Young, John R. B., Blacksmith

Middlesex Artillery

Fleet, William C., Captain
Hardy, William B., 1st Lt.
O'Brien, Robert, 1st Lt.
Roane, Timothy, 1st Lt..
Bristow, Lemuel C., Sr. 2nd Lt.
Keiningham, James, 2nd Lt.
Sutton, John W., 2nd Lt.
Seward, Edward W., Jr. 2nd Lt.
Montague, Christmas D., 1st Sgt.
Street, Robert B., 1st Sgt.
Chandler, John J., 2nd Sgt.
Beazley, George Philip, Sgt.
Bristow, William C., Sgt.
Roane, Addison M., Sgt.
Seward, Edmund W., Sgt.
Claybrook, William L., Sgt.
Zills, John, QM Sgt.
Keingingham, Gideon, Cpl.
Ailworth, Edmund, Cpl.
Revere, Elias Henry, Cpl.
Saunders, William J., Cpl.
South, Andrew J., Cpl.
Proffitt, William H., 5th Cpl.
Perry, Harvey D., 7th Cpl.
Rilee, Gideon R., Artificer
Dunn, Thomas H., Drummer

Gardner, Alexander Samuel, Drummer
Burrows, Mason M., Musician
Atkinson, Richard (or Robert) L., Bugler
Saunders, John R., Bugler
Saunders, John R., Fifer

Hanover Artillery

Nelson, William, Captain
Nelson, George Washington, Capt.
Cooke, Charles, 1st Lt.
Fontaine, William M., 1st Lt.
Terrell, Lewis F., 1st Lt.
Stringfellow, H. Martyn, Jr. 1st Lt.
Basye, William, 2nd Lt.
Points, Leonidas, Sgt. Maj.
Graves, Bernard B., 1st Sgt.
Wilson, W. A., Ord Sgt.
Holloway, Alexander J., Sgt.
Rouzie, Edward J., Sgt.
Southard, James T., Sgt.
Thacker, Andrew Jackson, Sgt.
Terrell, Nicholas, 3rd Sgt.
Stringfellow, Robert R., 4th Sgt.
Hall, Franklin S., 1st Cpl.
Flippo, William S., Cpl.
Grimes, George Washington, Cpl.
Nuckolls, Alpheus B., Cpl.
Taylor, Edmond, 2nd Cpl.
O'Brien, Philip, Musician

Magruder Light Artillery

Page, Thomas Jefferson, Captain
Magruder, George Allen, 1st Lt.
Smith, John Donnell, 1st Lt.
Yancey, Stephen D., 2nd Lt.
Bledsoe, Robert H., 1st Sgt.
Blum, Ernest, Sgt.
Brown, Dixon, Sgt.
Howard, Thomas C., Sgt.
Jones, Henry F., Sgt.
McKay, John, Sgt.
Herring, B., Cpl.
Bonhomme, Jean, Cpl.
Driscoll, Daniel J., Cpl.

Miles, Richard, Cpl.
Moore, William R., Cpl.
Newman, Thomas, Cpl.
Sherwell, William, Cpl.
Pitzenberger, Vincent, Bugler

Manchester Artillery

Weisiger, Emmett Washington, Captain
Weisiger, William Robert, Capt.
Gibbs, Thomas E., 1st Lt.
Abbott, James T., 2nd Lt.
Eubanks, Nathan A., 2nd Lt.
Perry, Artemus H., 2nd Lt.
Weisiger, Benjamin B., 2nd Lt.
Cox, Joseph Edwin, 3rd Lt.
Vaden, James A., 1st Sgt.
Montgomery, Benjamin D., 2nd Sgt.
Perry, Cyrenius T., 3rd Sgt.
Roundtree, Alexander W., 3rd Sgt.
Foster, John H., QM Sgt.
Holt, William H., 1st Cpl.
Bowman, William A. D., 2nd Cpl.
Hall, John W., 3rd Cpl.
Bowman, Robert C., 4th Cpl.
Cheatham, James W., 4th Cpl.
Hutchinson, William Henry, Bugler
Kelly, Spencer, Musician

Campbell Long Island Artillery

Clark, Patrick Henry, Captain
Wimbish, Abram, Capt.
Spencer, Thomas P., Jr. 1st Lt.
Hancock, N. M., 2nd Lt.
Williamson, Elbert M., 1st Sgt.
Scott, J. A., 2nd Sgt.
Connelley, William J., Sgt.
Jordan, Christopher Columbus, Sgt.
Macon, W. A., Sgt.
Mason, William A., Sgt.
Noell, Joseph R., 4th Sgt.
Barnes, James F., QM Sgt.
McGraw, Thomas G., Commissary Sgt.
Kabler, Thomas N., 1st Cpl.
Oliver, John Alonzo H., 3rd Cpl.

Ramsay, William A., 4th Cpl.
Wood, James B., 2nd Cpl.

Statistical Summary of Strengths and Losses

Note: The number of ages, occupations, residences varies depending on information provided in the CSR. The information provided herein is therefore not a complete accounting of statistics.

Washington/Hampton Artillery Company A, 1st Regt. Va. Art.

CATEGORY	Number
Wounded	1
Captured	1
Died of Disease	1
Transferred to Co. B, 1st Art.	2
Transferred to Co. F, 1st Art.	13
Transferred to King William Art.	5
Transferred to Co. B, 3rd Va. Cav.	2
Transferred to 32nd Va. Inf.	1
Elected/assigned to other co.	2
Discharged	6
Discharged & joined CS Navy	1
Appointed surgeon to another unit	1

*Fewer than 7 ages available. Only three listed occupations in the CSR. Birthplaces/residences at enlistment listed included Elizabeth City Co., Gloucester, Southampton Co., Henrico Co.

Beauregard Rifles
Company C (1st organization), 1st Regt. Va. Art.

CATEGORY	Number
Transferred to 1st Va. Cavalry	1
Discharged	8
Appointed P.A.C.S.	2
Commissioned into other company	1

Henrico Artillery
Company C (2nd organization), 1st Regt. Va. Art.

CATEGORY	Number
Died of Disease	1
Transferred to Co. C (2nd), 1st Art.	1
Transferred to Co. C (3rd), 1st Art.	42
Transferred to Co. H, 1st Art.	14
CM & NFR	1
CM & sent to Ft Caswell, NC	1
Discharged	21

AWOL & NFR	2
Deserted	11

Ages at Enlistment
Limited Number of Ages Available

17 - 2 (John Leathers & Frederick Seimms)
18 - 7
19 - 6

20 - 2
21 - 6
22 - 10
23 - 1
24 - 8
25 - 3
26 - 3
27 - 3
28 - 6

30 - 2
32 - 1
33 - 2
34 - 1
35 - 2
36 - 1
37 - 1
39 - 2

40 - 1
42 - 4
43 - 1
47 - 1 (Timothy Sullivan)

Occupations

blacksmith - 1
butcher - 1
cabinet maker - 2
carpenter - 8
coach maker - 1
cooper - 1
engineer - 3
engineman - 1
farmer - 5

finisher - 1
house joiner - 1
laborer - 17
lawyer - 3
machinist - 1
merchant - 2
molder - 1
oculist - 1
physician - 1
printer - 1
roofer - 1
sailor - 2
saw maker - 1
shoemaker - 2
tailor - 1
teacher - 2
teamster - 5
tinner - 1
tradesman - 1

Residences at Enlistment/Birthplaces

England - 1
Ireland - 5
Goochland Co. - 1
Hanover Co. - 1
Matthews Co. - 1
Nanesmond Co. - 1
New Kent Co. - 2
Prince Georges Co., Md. - 1
Richmond - 1
Wytheville - 1

Lee/Williamsburg Artillery
Company F, 1st Regt. Va. Art.

CATEGORY	Number
Captured	1
Died of Disease	3
Transferred to Co. D, 1st Art.	17
Transferred to Co. K, 1st Art.	18
Transferred to Powhatan Art.	3
Transferred to Medical Dept.	1
Discharged	8
Reassigned	1
Deserted	7

Peninsula Artillery
Company G, 1st Regt. Va. Art.

CATEGORY	Number
Wounded	1
Captured	2
Discharged	20
Died of disease	7
Dismissed	1
Detached	1
Exempted	1
Transferred to Morris Artillery	4

Ages at Enlistment
Limited Number of Ages Available

16 - 2 (Jeremiah Brand & William Sherwell)

21 - 1
25 - 1

34 - 1
35 - 3
36 - 2
38 - 2
39 - 1

40 - 1
43 - 2
45 - 1

* Few occupations listed including blacksmith (1), boatman (3), farmer (6), medical student (1), merchant (2), saddler (1), sailor (1), tailor (1),

** Few residences at enlistment listed including Gloucester (1), Matthews Co. (2), Nashville, Tennessee (1), Williamsburg (1), York Co. (8).

Wise Artillery

CATEGORY	Number
KIA	2
Wounded	8
Died of wounds	1

Died of Disease	4
Transferred to Bedford Artillery	59
Transferred to Eubank's Battery	1
Transferred to Maryland Line	8
Transferred to other batteries	1
Discharged	25
Retired	1
Deserted	3
Dropped from rolls	8
Resigned	1

Ages at Enlistment
Limited Number of Ages Available
14 - 1 (J.P. Mulligan)
16 - 1
18 - 2
19 - 1

21 - 1
22 - 3
24 - 1
29 - 2

31 - 2
32 - 2
33 - 1
34 - 2
36 - 1
37 - 2
38 - 3
39 - 1

41 - 2
42 - 2
43 - 1
44 - 1
46 - 1

51 - 1

Middlesex Artillery

CATEGORY	Number
Wounded & Captured	1
Died of Disease	2

Discharged	9
Transferred to Amherst Art.	13
Transferred to Jackson Flying	46
Transferred to Ashland Art.	18
Transferred to C.S. Navy	2
Transferred to Co. K, 55th Va. Inf.	1
Resigned	1
Dropped from rolls	1

Ages at Enlistment
Limited Ages Available

18 - 2
19 - 2

20 - 2
21 - 1
23 - 1
24 - 1

31 - 1
37 - 1

41 - 1
45 - 1
47 - 1

Residences included Albemarle Co. (1), Amelia Co. (1), Buckingham Co. (1),

Caroline Co. (1), Essex Co. (3), Fluvanna Co. (1), Gloucester Co. (3), Henrico Co. (1), King & Queen Co. (1), Middlesex Co. (21), New Kent Co. (1), Rockingham Co. (1).

Hanover Artillery

CATEGORY	Number
Died of disease	5
Discharged	9
Resigned	1
Relieved	1
Deserted	1
Transferred to Amherst Art.	22
Transferred to Ashland Art.	3
Transferred to Graham's Battery	1
Transferred to Parker's Battery	1

Transferred to Pollick's Battery	1
Transferred to James City Art.	1
Transferred to Co. G, 4th Va. Cav.	1
Transferred to 9th Va. Cav.	1

* Residences at enlistment included Ashland (50), Caroline Co. (3), Frederick Co. (1), Hanover Co. (40), Louisa Co. (8), Lunenberg Co. (1), Missouri (1), Spotsylvania Co. (1), Washington, D.C. (1). Also two (2) born in Ireland.

** Occupations at enlistment included blacksmith (1), carpenter (1), farmer (28), gentleman (1), lawyer (3), mechanic (12), medical doctor (1), merchant (1), operative mason (1), student (7), teacher (4)

Ages at Enlistment
Limited Number of Ages Available

18 - 6
19 - 8

20 - 6
21 - 6
22 - 5
23 - 3
24 - 9

25 - 4
26 - 2
27 - 2
28 - 4
29 - 1

30 - 2
31 - 1
33 - 1
34 - 1
36 - 1

41 - 1
42 - 2

52 - 1

Magruder Artillery

CATEGORY	Number

WIA	1
Deserted	1
Transferred to Bedford Art.	47
Transferred to Morris Art.	1
Transferred to C.S.S. Virginia.	1

* Residences of veterans included Baltimore, Md. (1), Charles City Co. (1), Franklin Co. (1), King & Queen Co. (1), New Kent Co. (1), New Orleans , LA. (9), Nottoway Co. (1), Warwick Co. (2), York Co. (5)

Manchester Artillery

CATEGORY	Number
KIA	2
Died of disease	3
MWIA	2
WIA	1
Discharged	12
Relieved	3
Resigned	2
Transferred to Courtney Art.	45
Transferred to Otey's Battery	1
Transferred to J.B. Jones Battery	1
Transferred to 39th Bttn. Va. Cav.	3
Transferred to 16th Va. Inf.	1
Transferred to Co. E, 44th Va. Inf.	1
Reassigned	1
Deserted	1
AWOL & NFR	13

Long Island Light Artillery

CATEGORY	Number
Wounded	1
Died of Disease	13
Discharged	5
Transferred to Orange Art.	21
Transferred to Morris Art.	14

* Residences of veterans included Bedford Co. (1), Campbell Co. (34), Charlotte Co. (8), Halifax Co. (1), Nelson Co. (1), Pittsylvania Co. (1).

John Hardy, Middlesex Artillery

Elbert Madison Williamson

George W. Walker, Wise Artillery

William R. Garrett
1st Virginia Artillery

A Note Regarding the Rosters

The rosters that follow were extracted from information in the Compiled Service Records, along with various postwar rosters updated by veterans. When available, birth, marriage, death and postwar dates and activities have been included. As far as the actual military records are concerned, all information has been copied with the exception of pay and in some instances clothing issue information. Sometimes it will be noted within the rosters that a veteran may have no more record available than "enl. 9/1/63. Pres. - 4/64." This usually suggests that either there was no more information available on the veteran from the service record or simply that nothing of particular note, personally, occurred. This is not to say that the veteran was not actively participating in movements, engagements, etc. Many Confederate records were lost at the close of the war due to various reasons. Sickness, hospitalizations, wounds, etc. have all been included in the rosters.

When dates are given in parenthesis, it will indicate that the dates are approximate based upon when an event was recorded in the service record. For example, when a person was described as AWOL "(10/31/64)" it simply shows that the date of the AWOL was noted in the muster rolls on 10/31/64, not that he went AWOL on that date.

Abbreviations Used In the CSR

ANV	Army of Northern of Virginia
Art.	Artillery
AWOL	Absent without leave
B.	Born
Battn.	Battalion
C.H.	Court House
Cav.	Cavalry
Cem.	Cemetery
Chimbo.	Chimborazo
CM or GCM	Court Martial/General Court Martial
Co.	County or Company as appropriate from context
comp.	complexion
Cpl.	Corporal
C.S.	Confederate States
D.	Died
Descr.	Described
DFR	Dropped from the Roll
Dstd	Deserted
Enl.	Enlisted or enrolled.
Exch.	Exchanged
Ft.	Fort

GH	General Hospital
GSW	Gunshot wound
G.O.	General Orders
Gen.	General
Hosp.	Hospital
HQ	Headquarters
Hvy	Heavy
Inf.	Inf.
KIA	Killed in Action
Lt.	Lieutenant or Light
Md.	Married
MIA	Missing in Action
mos	months
MWIA	Mortally wounded in action
NFR	No further record
occ.	occupation
Ord.	Ordnance
p.l.	Pension list
Pt.	Point
POW	Captured/Prisoner of War
Pres.	Present
Pvt.	Private
PWR	Post War Roster or Record
QM	Quartermaster
Res.	Resident of
S.O.	Special Orders
Sgt.	Sergeant
Surg. Cert.	Surgeon's Certificate
UCV	United Confederate Veterans (a veterans organization)
UDC	United Daughters of the Confederacy
U.S.	United States
UVA	University of Virginia
VMI	Virginia Military Institute
Vets	Veterans
WIA	Wounded in Action
yrs.	years

PRISONER OF WAR CAMPS

Camp Chase	Camp Chase, Ohio
Camp Morton	Camp Morton, Indiana
Elmira	Elmira, New York
Fort Delaware	Fort Delaware, Delaware
Fort McHenry	Fort McHenry, Maryland
Johnson's Island	Johnson's Island, Ohio
Newport News	Newport News, Virginia

Old Capitol	Old Capitol Prison, Washington, D.C.
Point Lookout	Point Lookout, Maryland
Rock Island	Rock Island, Illinois
Wheeling	Wheeling, West Virginia

Rosters of:
Co. A, Washington Artillery
Co. C-1, Beauregard Rifles
Co. C-2, Henrico Artillery
Co. F, Lee Artillery
Co. G, Peninsula Artillery
All of the 1st Regiment Virginia Artillery

ADAMS, HENRY: Co. G. Pvt. Enl. 9/20/61 at Tettlers Neck. In hospital in Williamsburg 9/22-10/31. Discharged disabled 11/20/61 at Williamsburg.

ADAMS, JOHN W.: Co. C-2. Pvt. Enl. 5/13/61, age 22; occupation: machinist. Sick in hospital (10/30/62). Deserted 12/1/62. Arrested as deserter and confined 4/7/63. Reported to Co. C- 3, 6/8/63.

ADAMS, WILLIAM D.: Co. A. Pvt. Bn. In Pennsylvania. Enl. 5/13/61 in Hampton, age 25. 5'8", dark comp., blue eyes, black hair, occupation: farmer. Discharged 6/24/62.

ADDISON, WILLIAM J.: Co. C-1. Pvt. Enl. 5/27/61, Manassas. Present thru (11/31/61). See 1st Virginia Inf.

ADLER, ABRAHAM: Co. A. QM Sgt. Enl. 5/13/61, Hampton. NFR.

ALLEY, THOMAS K.: Co. A, B. D. 9/19/64, buried in the Woodlawn Cem., Elmira, NY.

ANDERSON, ARTHUR H.: Co. C-1. 3rd Sgt. Enl. 5/1/61, Manassas. NFR.

ANDREWS, JOHN A.: Co. C-2. Pvt. Enl. 2/22/62, Richmond. Transferred to C-3.

ASKEW, HENRY S.: Co. C-2. Pvt. Born in Cork Co., Ireland. Enl. 5/13/61, age 45, merchant. AWOL (7/15/62). Discharged "over 35", 10/18/62. Age 46, 5'5", dark comp., black eyes, grey hair.

BADKINS, JOHN S.: Co. G. Pvt. Born in Williamsburg, Va. Enl. 6/1/61, Williamsburg. Discharged "over 35" 7/13/62. Age 38, 5'4", dark comp., blue eyes, black hair, tailor.

BAILEY, ROBERT: Co. C-2. Pvt. Enl. 5/13/61, Richmond. Age 25, laborer. AWOL (7/15/62). Present (8/31/62). Transferred to C-3.

BALLARD, JOHN P.: Co. C-2. Pvt. Enl. 5/13/61, Richmond. Age 18, farmer. Transferred to C-3.

BAMBERRY, HENRY: Co. C-2. Pvt. Enl. 5/13/61, Richmond. Absent at medical college in Richmond 5/30/61 due to fall from third story of the Baptist College. Fractured eye and broke thigh. Discharged.

BARNEY, PETER: Co. C-2. Pvt. Enl. 5/13/61, Richmond; age 28, laborer. AWOL (7/15/62). Deserter.

BARRY, ARTHUR R.: Co. C-2. Pvt. Bn. In Prince Georges Co., Md. Enl. 5/1/61 at Manassas; age 22, 5'8", fair comp., blue eyes, light hair, physician. Detached to serve in hospital in Charlottesville. Discharged 11/6/61. Commissioned asst. surgeon.

BARRY, WILLIAM D.: Co. C-1. Pvt. Enl. 5/1/61, Manassas. Present through (11/61).

BARTLETT, GEORGE: Co. F. Pvt. Enl. 5/20/61, Williamsburg. Transferred to Dance's Co.

BASSETT, THOMAS: Co. G. Pvt. Enl. 5/18/61, Upper Grafton. Absent at home in York Co. W/surgeon's leave from 2/18/62. Present (4/30). Listed as captured at battle of Williamsburg and sent to Ft. Monroe. NFR.

BAYLEY, WILLIAM H.C.: Co. C-1. 4th Sgt. Enl. 5/1/61, Manassas. NFR.

BELVIN, JAMES W.: Co. G. Pvt. Bn. Yorktown. Enl. 5/18/61, Upper Grafton; age 21, 6', fair comp., blue eyes, auburn hair, medical student. Improperly detailed as nurse in Richmond hospital. Absent from 9/1 - 10/31/61. AWOL 3/1/62. Discharged on surgeon's certificate as disabled 11/7/61.

BENNETT, JOHN P.: Co. C-2. Pvt. Bn. In Hanover Co. Enl. 5/13/61, Richmond, age 40, carpenter. Discharged 8/13/62 by conscription act; age 41, 5'7", dark eyes and comp., grey eyes.

BENTHALL, JOHN: Co. A, F, & D. Pvt. Enl. 5/13/61, Hampton. Transferred to Coke's Co. 7/21/62. Transferred from Coke's Co. To Co. D 10/6/62.

BENTON, ASA T.: Co. A. Pvt. Enl. 5/13/61, Hampton. Died 5/20/62 in Hospital #20 of typhoid fever. Effects: 1 silver watch.

BERRY, ENOCH R.: Co. C-1. Pvt. Enl. 5/1/61, Manassas. NFR.

BLAND, WILLIAM C.: Co. C-2. Pvt. Enl. 5/13/61, Richmond; age 21, carpenter. Transferred to C-3. Died in Sussex Co. 7/2/63, buried in the Blandford Cem., Petersburg, VA.

BLANKS, GLENDOR: Co. F & D. Enl. 6/10/62 near Richmond (Poe's Farm). Transferred 10/6/62 to Co. D.

BOISSEAU, JAMES: Co. C-2. Pvt. Enl. 5/13/61, Richmond; age 24, engineer. Detached to Engineer Corps 4/1/62. Present with Co. C-3 (10/31/63). Entered R. E. Lee Camp Soldiers Home, Richmond, 11/21/96, age 66. Died 9/19/98. Buried Hollywood Cem., Richmond, Va

BOONE, JOHN F.: Co. C-1. 1st Cpl. Enl. 5/1/61, Manassas. Absent sick w/typhoid in Richmond (11/61). Discharged 11/17/61.

BOSWELL, JAMES P.: Co. C-2. Pvt. Enl. 5/13/61, Richmond; age 42, cabinet maker. AWOL (7/15/62). Absent sick (8/31/62). Deserter. Discharged 5/2/63.

BOWIE, WALTER: Co. C-1. Pvt. Enl. 5/1/61, Manassas. Appointed captain in P.A.C.S. 9/23/61.

BOWLER, NEBUCHADNEZZAR: Co. C-2. Pvt. Enl. 5/13/61, Richmond. In arrest in Petersburg jail 12/9/62. Transferred to Co. C-3.

BOWLING, CONSTANTINE A.: Co. C-1. Pvt. Enl. 5/1/61, Manassas.

BOYLE, JOHN H., Jr.: Co. C-1. Pvt. Enl. 6/1/61, Manassas. Detached in Richmond to procure clothing for the company. Present (11/13/61).

BRAND, JEREMIAH: Co. G. Pvt. Bn. York Co. Enl. 6/15/61, Williamsburg. At G.H. #21, Richmond w/bilious fever 8/21/62. Transferred to Mayo Island 8/30. Discharged 10/8/62; age 17, 5'7", ruddy comp., grey eyes, lt. hair, boatman.

BREEDEN, BARTLETT B.: Co. C-2/3. Pvt. Enl. 5/13/61, Richmond; age 28, butcher. Transferred to C-3. Entered R. E. Lee Camp Soldiers Home, Richmond, 12/7/96, age 69. Dropped 2/16/97.

BREEDEN, RUFUS: Co. C-2/3. Pvt. Enl. 5/13/61, Richmond; age 22, laborer. Absent sick 10/61. Arrested near Richmond as deserter; charged $30. Present (4/30/62). In arrest (7/15/62). Present (8/31/62). AWOL (10/31/62). Deserted 9/1/62. Transferred to C-3, 2/15/63 after returning from desertion.

BRENT, ALEXANDER M.: Co. C-1. Pvt. Enl. 5/1/61, Manassas. NFR.

BRIDGEWATER, PATRICK HENRY: Co. C-2/3. Pvt. Enl. 5/13/61, Richmond; age 30, teamster. Transferred to C-3.

BRIDGEWATER, ROBERT: Co. C-2/3. Pvt. Enl. 5/13/61, Richmond; age 26, teamster. Transferred to C- 3.

BRIGHT, ROBERT A.: Co. G. 1st Lt. Enl. 5/18/61, Upper Grafton. 2nd Lt. (7/1/61). Listed as brevet 2nd Lt. 10/13/61. Absent on leave 4 days, 12/30/61 in Williamsburg. Detailed to take charge of heavy baggage at Lebanon Church 2/4/62. 3rd Lt.?? 2/28/62. Present (4/30/62). At G.H. #21, Richmond w/chronic diarrhea and later transferred to private quarters in Amelia Springs 5/23/62. 1st Lt. 8/9/62. See also General and Staff, 53rd Va. Inf. Captain on staff of Gen. G.E. Picket. Surrendered at Appomattox.

BROOKE, EUGENE: Co. C-1. Pvt. Enl. 5/1/61, Manassas. Absent sick in Richmond (11/13/61). Deserted? NFR.

BROOKE, THOMAS V.: Co. C-1. Pvt. Enl. 5/23/61, Manassas. Transferred to Co. D.

BROOKE, WALTER B.: Co. C-1. Pvt. Enl.6/19/61, Manassas. Sick in Charlottesville hospital (11/13/61). At Chimbo. #4, 11/27/61 with fever. Transferred to Robertson Hospital, 12/25/61. NFR.

BROWN, DIXON: Co. G. 1st Cpl. Enl. 5/18/61, Upper Grafton. Present through (4/30/62).

BROWN, WASHINGTON A.: Co. C-2/3. Pvt. Bn. In Chesterfield. Enl. 5/13/61, Young's Mill; age 26, cooper. Reenlisted 2/28/62 in Warwick Co. Transferred to C-3. 5'10", light comp. and hair, blue eyes.

BRUGH, WILLIAM P.: Co. C-2 & H. Pvt. Enl. 3/10/62, Fincastle. Transferred from Ewing's Battery 6/26. Transferred from Ritter's Battery to Co. H, 10/28/62.

BUCHANAN, HENRY W.: Co. A. Pvt. Enl. 5/13/61, Hampton. Present through (10/31/61). NFR.

BUCHANAN, MARTIN: Co. F & K. Pvt. Enl. 5/20/61, Williamsburg. Transferred to Co. K, 10/6/62.

BUCHANAN, WILLIAM: Co. F & K. Pvt. Enl. 5/20/61, Williamsburg. Absent sick in hospital (5/26/62). Present (7/12/62). Transferred to Co. K, 10/6/62.

BULLOCK, MATHEW B.: Co. C-2/3. Pvt. Enl. 2/20/62, Richmond. Absent sick with leave (7/15/62). Present (8/31/62). Transferred to C-3.

BULLY, NATHANIEL: Co. A. Pvt. Enl. 6/13/61, Yorktown. Present through (4/30/62). NFR.

BUNCH, McWILLIAM: Co. A. Pvt. No record while serving w/Co. A officially. Only noted as captured 8/10/63 at Farmville, Va. and at Old Capitol Prison, Washington, D.C. Sent to Kalorama, 11/23/63. D. 12/3/63 of variola; age 20. Buried at 4 p.m., 12/4. Residence before enlisting: Springfield, Illinois.

BURNS, NICHOLAS: Co. C-2. Pvt. Enl. 5/13/61, Richmond. Deserter 5/13/61. Resident of Richmond. NFR.

BURROWS, HOWARD L.: Co. I & C-2. Ord. Sgt. Enl. 4/25/61; age 19, teacher. Discharged 6/25/61. Left 6/21 for Richmond under 5 day furlough. Transferred to Sands' Company. Reduced to ranks 6/30/61. NFR.

BUSH, CHARLES E.: Co. C-2/3. Pvt. Enl. 3/10/61, Fincastle. Transferred from D.P. Ewing's Company. Absent sick on leave (8/13/62. Deserted 7/1/62.

CAINE, GEORGE WASHINGTON: Co. A & B. Pvt. B. 4/14/38. Enl. 5/13/61, Hampton. At some point transferred to Co. B. Transferred to Carter's Battery. WIA at Gaines Mill, VA in 1862, incapacitated for further service. Discharged as disabled 12/30/62. Carpenter postwar in Hampton. Member of the Greble Lodge Odd Fellows; member R. E. Lee Camp Confederate Veterans; member 1st Methodist Church of Hampton. D. 3/3/1920 and buried in the Odd Fellows Cem., Hampton, VA.

CAKE, RICHARD B.: Co. A. 5th Sgt. Enl. 5/13/61, Hampton. 2nd Cpl. (12/31/61). 1st Cpl. (2/28/62). 5th Sgt 4/12/62. Transferred to Carter's battery. Resident of Elizabeth City County. Alive in 1920.

CALLAGHAN, FRANCIS: Co. C-2 & H. Pvt. Enl. 8/12/62, Petersburg. Transferred from Ritter's Co. To Co. H, 10/28/62.

CALLAS, DANIEL J.: Co. G. Pvt. Enl. 6/15/61, Williamsburg. At G.H. #21, Richmond 6/28/62 w/typhoid fever. Returned to duty 9/3/62. Transferred to Page's Battery

CARDWELL, BERRY: Co. F. Pvt. Enl. 5/20/61, Williamsburg. Sick in hospital (12/31/61). Discharged 1/28/62.

CARDWELL, PATRICK HENRY: Co. F & D. Pvt. Enl. 5/20/61, Williamsburg. Transferred to Co. D, 10/6/62.

CARPER, ANDREW JACKSON: Co. C-2. Pvt. Enl. 3/10/62, Fincastle. Absent sick (7/15/62). Transferred from Ewing's Battery. Transferred to Co. H, 10/28/62.

CARUTHERS, ROBERT: Co. F. 3rd Cpl. Enl. 5/20/61, Williamsburg. 3rd Cpl. 5/10/62. Captured at Antietam and took oath 9/27/62 near Sharpsburg, Maryland.

CARVER, MOSES MONROE: Co. A. Postwar reference. B. 4/6/1837 D. 10/18/1909, buried St. John's Cem., Hampton, VA.

CASEY, JOHN WESLEY: Co. F & K. Pvt. Enl. 5/20/61, Williamsburg. Transferred to Co. K, 10/6/62.

CHAPMAN, JOHN F.: Co. G. 3rd Cpl. B. in York Co. Enl. 5/18/61, Upper Grafton. Absent on surgeon's leave 11/6 at home in York Co. Present (2/28/62). At G.H. #21 w/hemorrhoids 5/9/62 5/27/62. Discharged 7/16/62; age 26, 5'6½", dark comp., grey eyes, black hair, farmer. Newport News p.l. Entered R.E. Lee Camp Soldiers Home, Richmond, 4/15/04, age 68. Dismissed 5/19/04.

CHARLES, HENRY H.: Co. F & K. Pvt. Enl. 5/20/61, Williamsburg. On detached service to guard baggage (12/31/61). Present (5/25/62). Transferred to Co. K, 10/6/62. MWIA, 12/13/62 at Fredericksburg. Died same night.

CHARLES, WILLIAM H.: Co. F. 3rd Cpl. Enl. 5/20/61, Williamsburg. Absent sick (5/26/62) - (9/30/62). NFR.

CHEATHAM, WILLIAM T.: Co. C-2. Pvt. Enl. 5/13/61, Richmond; age 43, farmer. In hospital as attendant, 9/14/61. Discharged as disabled at Young's Mill, 12/7/61 by order B. Gen. McLaws on surgeon's cert.

CLARKE, ROBERT J.: Co. C-2. Pvt. Enl. 5/13/61, Richmond; age 25, carpenter. Furloughed 6/17/61 to Essex Co. Severely sick. Discharged 12/5/61, Richmond on surgeon's certificate of disability.

COFFIN, NORRIS W.: Co. C-1. Pvt. Enl. 5/1/61, Manassas. See Coffin's Battery.

COKE, JOHN ARCHER: Captain. Co. F. Captain. B. 7/14/42, Williamsburg, Va. Attended College of William and Mary, A.B., 1860. Enl. 5/20/61 in the Lee Artillery. Elected 2nd lieutenant, (12/31/61). Captain, 5/62. In G.H. #8 (St. Charles Hosp.), Richmond and "gone to home", 8/3/62. When the battery was disbanded, Coke was ordered to report to General G.W. Smith, at Richmond, and assigned duty as enrolling officer for the Third Congressional district of Virginia, 5/6/64. Volunteered in every opportunity to participated in repulsing the enemy from Richmond. While aide-de-camp during one action, Coke was wounded in the head by a piece of shell, but not seriously injured. When Richmond was evacuated, he accompanied General J.L. Kemper, as a member of his staff, with a party of officers to Halifax Co. where they were met with the news of Lee's surrender. Proceeded to Henry Co. and joined a party that was going to join Joseph E. Johnston's army in North Carolina. Upon hearing of Johnston's surrender, returned to Danville and gave his parole. Returned home to begin study of law. 9/65, moved to Richmond and entered into a practice. Member of R.E. Lee Camp No. 1, Confederate Veterans. D. 1/27/20, Richmond, Va., age 77.

COLBERT, MARION: Co. A. Pvt. Enl. 5/13/61, Hampton. In Richmond on 20 day sick furlough 12/6/61. sent sick in Richmond (2/25/62). Detailed 3/31/62. Discharged 7/29/62 per S.O. #75.

COLE, FREDRICK: Co. A. Pvt. Enl. 6/5/61, Yorktown. Absent sick in York Co., (2/28/62) since 12/6 on surgeon's certificate.

COLLINS, JOHN: Co. C-2. Pvt. Enl. 5/13/61, Richmond. Sent to hospital 4/1/62. At Danville G.H. w/acute diarrhea 5/6/62. Returned to duty 7/30/62. Discharged 8/13/62 by conscription act. Captured at Guinea Station 6/11/64. Sent to Pt. Lookout and released 6/27/64 after joining U.S. service.

COLLONY, JAMES E.: Co. A. Pvt. Enl. 5/13/61, Hampton. NFR.

COOK, ANDERSON: Co. F. Pvt. Enl. 5/20/61, Williamsburg. At Williamsburg Hospital w/fever 9/11 - 9/15/61. Transferred to Dance's Company.

COOK, CHARLES: Co. C-2. Pvt. Enl. 5/13/61, Richmond; age 22, teamster. Discharged on surgeon's certificate as disabled, 9/21/61.

COOPER, ELIJAH A.: Co. A. Pvt. Enl. 5/13/61, Hampton. Detailed as wood cutter for post, 1/62. Transferred to Carter's Co. Alive in 1920.

COSNAHAN, JOSEPH B.: Co. G. Captain. From South Carolina. Attended William & Mary College & UVA. Enl. 5/18/61, Upper Grafton. Absent on leave in Williamsburg, 4/29/62. Died in service, 1862. Widow (Ellen J. Cosnahan) filed claim.

COURTNEY, ALFRED RANSON: Lt. B. 11/17/33, King & Queen Co., Va. Enl. with the Hampton Artillery and became lieutenant. Elected captain of a battery from Henrico County about 7/8/61, which later bore his name as the Courtney Artillery. Major of artillery, chief of artillery, Ewell's Division, 7/14/62. Relieved of duty in the A.N.Va. at his own request, and assigned to A.I.G.O., 4/20/63. He had been court-martialed for dereliction of duty at Sharpsburg, having failed to go into battle with the batteries, and found guilty on most counts. Transferred to the Dept. of East Tennessee, Chief of Artillery in Hood's Division, Army of the Tennessee. Seriously wounded at the battle of Resaca, Ga. Spring, 1865, put in command of a battalion of artillery organized from the remnants of the Tennessee campaigns and reported to General P.G.T. Beauregard at Augusta, Ga. Paroled in Augusta, Ga. Returned to Richmond and began career as lawyer. Elected Commonwealth Attorney for Henrico Co., but was debarred from holding the post by the military authority. In 1870, he was elected to the State Senate from Henrico Co. and served one term. Grand Master of Virginia Masons. D. 11/4/14, Richmond, Va.

COX, GEORGE THOMAS: Co. C-1. 2nd Lt. Enl. 5/1/61, Manassas. Absent on detail (11/61). NFR.

CRANDALL, CHARLES P.: Co. C-1. Pvt. Enl. 7/1/61, Camp Stevens. NFR.

CRANDALL, WILLIAM A.: Co. A, F & K. Pvt. Enl. 5/13/61, Hampton. Transferred to Coke's Battery, 7/21/62 from Smith's Battery. Assigned to Co. K, 10/6/62.

CRANFORD, GEORGE J.: Co. C-1. Pvt. Enl. 5/1/61, Manassas. NFR.

CRAVER, MONROE MOSES: Co. A. Pvt. Enl. 5/13/61, Hampton. Transferred to Carter's King William Artillery, 6/14/62. Furlough 1/64 - 2/29/64. POW at Spotsylvania C.H. 5/12/64. Sent to Belle Plain, then to Ft. Delaware 5/21/64. Exchanged 10/30/64 at Savannah, Ga. & admitted to Richmond hospital 11/64. Joined the Richmond Howitzers but not on CSR. Assigned to heavy artillery at Petersburg. Paroled in Richmond 4/22/65. Illiterate; res.: Baltimore. Living at 218 S. Charles St., Baltimore in 1894.

CRAWLEY, WALLACE B.: Co. C-2. Enl. 2/20/62, Richmond. Transferred to C-3.

CRAY, WILLIAM R.: Co. C-2. Pvt. Enl. 5/13/61, Richmond; age 18, printer. AWOL (7/15/62). Transferred to C-3.

CROCKWELL, JOHN R.: Co. C-1. Pvt. Enl. 5/1/61, Manassas. NFR.

CROPPER, GEORGE T.: Co. A, F & D. Pvt. Enl. 5/13/61, Hampton. Transferred to Coke's Co., 7/21/62. Transferred to Co. D, 10/6/62.

CRUMP, SYLVANUS G. : Co. C-2. Jr. 1st Lt. B. New Kent Co. Enl. 5/13/61, Richmond; age 24. Cpl. Sgt (2/28/62). Absent on furlough (2/62) - 3/16/62. Cpl. (4/30/62). Jr. 1st Lt. (7/15/62) in place of W. H. Robertson. Signed as commanding officer. 6'1½", age 25, dark comp., hair and eyes.

CRUMP, WILLIAM B.: Co. C-2. 2nd Lt. B. New Kent Co. Enl. 5/13/61 Richmond as Sgt. Age 27, 6'1½", dark comp., hair and eyes. In hospital in Williamsburg, 12/61. Furloughed 3/16/62. 2nd Lt. in place of W.B. Ritter, 4/30/62. At Petersburg G.H. w/gonorrhea, 5/11 - 5/31/63. Deserter and came into U.S. lines at Williamsburg, 7/3/64. Released, 7/10/64; light comp., black hair, and brown eyes, 6'3".

CRUSH, WILLIAM L.: Co. C-2 & H. Pvt. Enl. 3/10/62, Fincastle by Captain E.F. Bowyer. Transferred 6/26 to C-2. Absent sick (7/15/62). On sick leave (8/31/62). AWOL (12/31/62).

CULLEY, WILLIAM A.: Co. F & K. Pvt. Enl. 5/20/61, Williamsburg. Assigned to Co. K, 10/6/62.

CUMMINS, EDWARD H.: Co. C-1. 1st Lt. Enl. 5/1/61, Manassas. Absent on detached service (11/61).

CUNNINGHAM, JAMES H.: Co. A. Pvt. No record of service. Enl. 4/24/64 w/Charlottesville Artillery at Liberty Mills. Paroled at Gordonsville, 5/24/65.

CURLEY, HENRY C.: Co. C-1. Pvt. Enl. 5/29/61, Manassas. NFR.

CURTIS, ROBERT or ROBERT KEITH: Co. A. B. 7/27/44 in Gloucester Co., Va. Son of Colonel Robert C. and Elizabeth Wray (daughter of Captain George Wray, of the Revolutionary Army). Educated at Cary's Military Academy at Hampton until he was fifteen, when he enlisted in the Washington Artillery in Hampton, 5/13/61. Detailed for special duty to surgeon 5/27/61/ Later assigned to the Old Dominion Dragoons, Co. B, 3rd Va. Cavalry, 10/18/61. Wounded four times simultaneously at Gettysburg, "fainting, fell from his horse, and would have been captured" had it not been for another comrade. Rode back to Gordonsville after the campaign where he finally received medical attention. Until recovered, he remained with his family in North Carolina, and then returned to serve as scout attached to the headquarters of General Fitzhugh Lee. Married in 1/27/72 to Margaret M. Sinclair, daughter of Fayette Sinclair, also a Confederate veteran. She died in 1895, leaving 8 surviving children. Farmer in Elizabeth City County after the war, until 1891 when elected sheriff, "the first Democrat since the war." Re-elected in 1895. Lt. Commander of the R.E. Lee Camp, UCV.

DAILEY, THOMAS: Co. C-2. Pvt. B. Nanesmond Co. Enl. 5/13/61, Richmond; age 42, laborer. AWOL (7/15/62). Entitled to discharge 8/13. Discharged 8/14/62, "over 35"; age 39, 5'10", dark comp. and eyes, brown hair, mason. Detailed to hospital in Richmond (8/31/62). At G.H. #22, Richmond until (2/63).

DAUGHERTY, WILLIAM T.: Co. A. 2nd Lt. (postwar ref). 3rd Sgt. Co. A. B. 12/13/40 in Hampton, Va. Son of George P. Daugherty of Southampton County, who was for many years the foreman of the Gosport Navy Yard. Enl. 5/13/61, Hampton. Furloughed at Yorktown, 12/29/61. 2nd Lt. Resigned commission on 5/20/62 to become a private in the Old Dominion Dragoons and served with them, losing three horses shot from under him, and was captured 8/16/64 in Front Royal. Spent a week in Old Capitol Prison, then transferred to Elmira until 3/10/65, when he accepted a parole for three months, with other invalid prisoners. After the war, became a carpenter and builder, his "principal work being the Hygeia hotel at Old Point Comfort" In 1866, married Mary L. Curtis of Hampton. Seven children. Elected commissioner of revenue in Hampton, 1895. Member of the board of trustees of the Hampton School. One of the charter members of the camp in Hampton and was the first commander. Adjutant of R.E. Lee Camp, No. 3, United Confederate Veterans, Hampton, Va.

DAVIS, DAVID N.: Co. C-2 & H. Pvt. Enl. 3/10/62, Fincastle. Transferred from Bowyer/Ewing's Co., 6/26/62. Transferred to Co. H.

DAVIS, JAMES: Co. A. 1st Cpl. Enl. 5/13/61, Hampton. 2nd Cpl. (2/25/62). 1st Cpl., 4/12/62. Furloughed to York Co., 12/29/61. NFR.

DAVIS, LARKIN W.: Co. G. Pvt. Enl. 5/18/61, Upper Grafton. Absent in hospital 10/21 - 10/31, Williamsburg. At home on surgeon's certificate, York Co., 10/31. Discharged as disabled 1/12/62.

DAVIS, MARSHALL J.: Co. G. Pvt. B. York Co. Enl. 5/18/61, Upper Grafton. In Seminary Hospital, Williamsburg with fever, 8/1 - 8/18/61. On leave for 30 days in York Co. from 1/22/62. Absent on leave for 4 days, 2/26/62. Present (4/30/62). At G.H. #21, Richmond with wound, 7/2/62. To private quarters, 7/4. Discharged 7/15/62; age 45, 5'11", dark comp., grey eyes, light hair, farmer.

DAVIS, SAMUEL H.: Co. A. & D. Pvt. Enl. 5/13/61, Hampton. Transferred to Coke's Co., 7/21/62. Transferred to Co. D, 10/6/62.

DAVIS, WILLIAM H.: Co. A. Pvt. B. Elizabeth City Co. Enl. 5/13/61, Hampton. On surgeon's certificate as furloughed in York Co., 12/6/61. Discharged as "under 18", 8/13/62; age 17, light comp., eyes, and hair, student.

DECOSSY, JAMES: Co. C-2. Pvt. B. Ireland. Enl. 5/13/61, Richmond; age 33, tradesman. Discharged 8/13/61; age 38, 5'9", dark comp. and hair, grey eyes.

DEMENT, THOMAS JEFFERSON: Co. C-1. Pvt. Enl. 5/1/61, Manassas. NFR.

DIGGS, JOHN M.: Co. C-2/3. Pvt. B. Matthews Co. Enl. 5/13/61, Richmond; age 22, sailor. AWOL (2/28/62). Present (4/30/62). Transferred to C-3. KIA 8/21/64. 5'8", light comp. and hair, blue eyes.

DIXON, L. W.: Co. C-2. Postwar reference. Died in service, buried on 12/18/64 in Laurel Grove Cem., Shenandoah Co., VA.

DODD, WILLIAM FLETCHER: Co. F. 5th Sgt. Enl. 5/20/61, Williamsburg. 5th Sgt., 5/10/62. NFR.

DODSON, RICHARD T.: Co. C-1. 3rd Sgt. Enl. 5/1/61, Manassas. Sick in Richmond, 11/13/61.

DODSON, THOMAS: Co. C-2/3. Pvt. Enl. 4/11/62, Richmond. Transferred to C-3.

DOHERTY, ANTONY: Co. C-2. Pvt. Enl. 5/13/61, Richmond; age 28, laborer. Absent sick w/leave (7/15/62). Transferred to Co. H, 10/28/62.

DOHERTY, CHARLES: Co. C-2. Pvt. Enl. 5/13/61, Richmond; age 19, sawmaker. Transferred to C-3.

DOHERTY, JOHN: Co. C-2/3. Pvt. Enl. 5/13/61, Richmond; age 22, carpenter. Transferred to C-3. 5'6", dark comp. and hair, blue eyes.

DOHERTY, PATRICK: Co. C-2. Pvt. Enl. 5/13/61, Richmond; age 26, sailor. AWOL (7/15/62). Discharged 7/23/62, not subject to conscription act.

DONLEY, CHARLES: Co. G. Pvt. Enl. 5/18/61, Upper Grafton. Absent on surgeon's leave 12/10 at York Co. Home. NFR.

DREW, JOHN WATERS: Co. C (Beauregard Rifles). Pvt. B. 1841, Washington, D.C. Attended Columbia College. Enl. 5/61 at Alexandria, Va. CSR lists him as joining Co. C-1, 7/1/61 at Camp Stevens. Commissioned captain of Co. F, 23rd Va. Cavalry, 10/1/63. Captured 9/19/64. Dark comp. and hair, gray eyes, 6'. Sent to Ft. Delaware. Paroled and returned to Washington, becoming a pharmacist. Member of the Washington Association of Confederate Veterans. D. 7/31/17; buried in Arlington Cem., Arlington, Va.

DUNCAN, JAMES P.: Co. C-2. Pvt. B. in England. Enl. 5/13/61, Richmond; age 40, tailor. At hospital sick (10/31/61). Present (2/28/62). Absent on sick leave (7/15/62). Discharged 8/13/62; light comp., blue eyes, grey hair.

EDELIN, WILLIAM H. H.: Co. C-1. Pvt. Enl. 5/27/61, Manassas. Absent sick in Richmond (11/13/61) at Chimbo. #5. Discharged 11/16/61, typhoid.

EDWARDS, JOHN: Co. F. Pvt. Enl. 3/18/62, Richmond. Transferred from Cropper's Co., 6/11/62. Deserted 8/20/62. Brought back under guard. Sentenced to be shot to death but pleaded not guilty - found guilty, 12/6/62. Apparently was not shot. Federal provost reported him as member of Co. D; age 51, 5'5", dark comp., blue eyes, sandy hair, farmer.

EDWARDS, RAYMOND D.: Co. C-1. Pvt. Enl. 7/1/61. Present through (11/13/61). NFR.

ELIASON, HENRY A.: Co. C-1. Pvt. Enl. 5/1/61, Manassas. Pres. Through (11/13/61). NFR.

ENROUGHTY, MADISON: Co. C-2/3 & U.S.A. Pvt. Enl. 5/13/61, Richmond; age 23, blacksmith. Sent to hospital. Deserted and under arrest (fined $30). Pres. (2/28/62). AWOL (8/31/62). Returned 2/15/63.

Transferred To C-3. Captured at Ft. Harrison, 9/30/64. Joined U.S. service, 10/15/64 from Pt. Lookout.

ENROUGHTY, NELSON: Co. C-2/3. Pvt. B. 6/2/1843. Enl. 5/13/61, Richmond; age 20, teamster. AWOL (7/15/62). Pres. (8/31/62). AWOL (10/31/62). Deserted 9/15/62. Returned 2/15/63. Transferred To C-3. At G.H. #9, Richmond, 2/25/65. D. 11/9/1917 in Richmond, VA.

ENROUGHTY, RICHARD: Co. C-2/3. Pvt. Enl. 5/13/61, Richmond; age 22, laborer. Transferred to C-3. Transferred from Young's Battery on 12/26/63 by S.O. 306, para 8. Pres. on 10/31/64 and 12/31/64 rolls. Paroled at Appomattox C.H., VA on 4/9/65.

ENROUGHTY, WALTER: Co. C-2/3. Pvt. Enl. 5/13/61, Richmond; age 27, teamster. In hospital after paying $30 for desertion. Pres. (2/28/62). Transferred To C-3.

EUBANK, GEORGE WASHINGTON: Co. C-2/3. Pvt. Enl. 5/13/61; age 24, carpenter. Absent sick (7/15/62). Deserted 10/1/62. Present in C-3 (6/30/63).

EVERETT, CHARLES D., M.D.: Co. G. Exempted from military service in 1861, furnished a substitute. Provided uniforms and arms for an entire company-the Albemarle-Everett Art. B. 1806 in KY. D. 1/1877 in Nelson Co., VA.

FAHERTY, WILLIAM J.: Co. C-1. Pvt. Enl. 5/1/61, Manassas. Absent sick in Richmond (11/13/61). Deserter? Sent to Orange C.H., 10/9/61, general hospital.

FILLINS, JOHN W.: Co. C-1. Pvt. Enl. 8/1/61, Fairfax C.H. Pres. Through (11/13/61). NFR.

FINN, WILLIAM: Co. C-2. Pvt. B. Ireland. Enl. 5/13/61, Richmond; age 34, laborer. Discharged 8/13/62; age 38, 5'10", dark comp., dark grey hair, grey eyes, carpenter.

FLIPPIN, MARTIN V.: Co. C-2. 1st Cpl. Enl. 5/13/61, Richmond; age 19, farmer. In hospital sick (2/28/62). Pres. (4/30/62). 1st Cpl., 7/1/62. Transferred To C-3.

FOSTER, JOHN: Co. G. Pvt. B. York Co. Enl. 5/18/61, Upper Grafton. Discharged 7/15/62; dark comp., hair, eyes, farmer. York p.l.

FRASER, WILLIAM W.: Co. A. Captain. Enl. 5/13/61, Hampton. Appointed 1st Lt., 7/22/61. Paroled 5/1/65, Winchester. Assigned per S.O. #125/6, 5/30/64.

FRENCH, CHARLES: Co. F. Pvt. Enl. 9/23/61, Williamsburg. Sick in hospital (12/31/61). Pres. (7/12/62). Discharged 8/21/62.

GAMBOL, ROBERT J.: Co. A & D. Enl. 5/13/61, Hampton. Transferred To Coke's Co., 7/23/62. Transferred To Co. D, 10/31/62. Captured at Spotsylvania C.H., 5/10/64. Sent to Ft. Delaware then to Pt. Lookout and exchanged 10/31/64.

GARNES, ALBERT: Co. C-2 & H. Cpl. Enl. 5/13/61, Richmond; age 22, house joiner. Transferred To Co. H, 10/28/62.

GARNES, WILLIAM H.: Co. C-2. Pvt. Enl. 5/13/61, Richmond; age 18. Court Martialed 6/3/61 at the Hermitage Fair Grounds. Violated 46th Article of War. Sleeping at his post #11, 10:00 p.m. at Artillery Barracks, Baptist College, 5/30/61. Pleaded guilty but claimed being sick for 4 days prior; "the Sgt. Told me I had to go on guard, I told him I was sick but he said I had never been in guard and I had better go then. I told my whole squad that I was sick. I went on guard and was taken __ that night. I have only been mustered into service three weeks today and have never had the articles of war read to me." Found guilty to specification and charge and was sentenced to hard labor for 30 days in the presence of the company w/ball and chain weighing six pounds attached to his legs. NFR.

GARRETT, WILLIAM ROBERTSON: Co. F. Captain. B. 4/12/39, Williamsburg, Va., son of Dr. Robert Major and Susan Winder Garrett. Father was mayor of Williamsburg and surrendered the city to McClellan, 5/61. Entered College of William and Mary, 1854. Graduated and recd. degree of A.M., 7/58. Attended U.Va. and studied law until war broke out. Enl. 4/61 in 32nd Va. Inf. Enl. 5/20/61 in Williamsburg. Elected captain of Lee Artillery. Absent on leave (12/31/61). After company disbanded, accepted a commission to raise a company of cavalry for partisan ranger service in Tennessee. In a postwar account, "At this juncture the crew of the Merrimac arrived in Richmond, and about 100 of them enlisted with Captain Garrett for that service. A naval officer, having received authority from the navy to organize a fleet of small boats to attack the Galena, which was near Drewry's Bluff, offered Captain Garrett command of one division of the canoe fleet. The proposition was accepted, and the boats, manned by the Merrimac men, went rapidly down the river to Drewry's Bluff. There they reported to General Mahone, only to find that the Galena had departed, and that an order from the war department had been issued at the request of the navy department to detain all the Merrimac crew as marines to man the guns at Drewry's Bluff and for future service in the navy, the claim being made that their term of enlistment in the navy had not expired. Captain Garrett returned to Richmond, and , a few days afterward, accompanied by Maj. D.C. Douglass and others, went to Tennessee for the purpose of recruiting a battalion of partisan rangers. The party entered the Federal lines in middle Tennessee and speedily organized Douglass' battalion of partisan rangers, of which Garrett was made adjutant." Upon the repeal of the partisan law, the battalion was mustered in as regular cavalry and consolidated with Holman's battalion, forming the 11th Tennessee Cavalry. Became a part of Dibrell's brigade of Nathan Bedford Forrest's Cavalry. Part urrendered at Gainesville, Alabama. Garrett had been adjutant and captain of Company B. After the war, he returned to Williamsburg to practice law but was offered a position as master of the grammar school of the college. In 1/68, he moved to Tennessee where he married Julia Flournoy Batte, of Pulaski, Giles Co. President of Giles College, superintendent of public instruction in Giles, principal of Cornersville Academy, professor of mathematics in Montgomery Bell Academy of the University of Nashville., State superintendent of public instruction, principal of the military academy at Nashville, president of the State Teacher's Association, president of the National educational association of the United States at the international meeting in Toronto, Canada; professor of American History at the Peabody Normal College, Nashville, Tennessee; editor of the American Historical Magazine; vice-president and chairman of the historical committee of the Tennessee association; member of the committee on history of the United Confederate Veterans; and one of the trustees of the Confederate memorial association. D. 2/12/04, Nashville, Tennessee.

GIBBS, JOSEPH: Co. C-2. Pvt. Enl. 3/10/62, Fincastle. Transferred from Ewing's Co. Absent sick (8/31/62). NFR.

GIDDINGS, WILLIAM: Co. A & F. Pvt. Enl. 5/13/61, Hampton. Detailed to Captain Archer at West Point, Va., 11/17/61. Pres. (4/30/62). Transfer to Coke's Co., 7/21/62.

GLEASON, THOMAS: Co. C-2. Pvt. B. Ireland. Enl. 5/13/61, Richmond; age 19, laborer. Discharged 7/20/62; age 17, 5'6", light comp. and hair, blue eyes. Listed as deserter 9/62.

GOVAN, CHARLES: Co. C-2/3. Pvt. Enl. 5/13/61, Richmond; age 22. Deserter 5/25/61. Present (4/30/62). Transferred To C-3.

GRADY, JOHN: Co. C-2/3. Pvt. Enl. 5/13/61, Richmond; age 30, laborer. AWOL (7/15/62). Deserted 4/1/62. Pres. And Transferred to C-3.

GRAHAM, JAMES: Co. C-2. Cpl. Enl. 5/13/61, Richmond. Cpl. (7/15/62). Died 8/10/1862. Buried in Hollywood Cem., Richmond, VA.

GRAVES, JOHN: Co. G. Pvt. Enl. 5/18/61, Upper Grafton. At G.H. #21, Richmond w/acute diarrhea, 5/9/62 - 5/13/62. Transferred To Page's Battery.

GRAVES, RICHARD: Co. G. Pvt. B. Williamsburg. Enl. 6/1/61, Williamsburg. Detailed as harness maker in QM dept., 9/1 - (2/28/62). Discharged 7/25/62; 5' 10", dark comp. and eyes, black hair, saddler.

GREEN, AUGUST: Co. F. Pvt. Enl. 3/18/62, Richmond. Transferred from Cropper's Battery, 6/10/62. Deserted 8/20/62.

GREEN, PETER: Co. G. Pvt. Enl. 5/18/61, Upper Grafton. Absent on surgeon's certificate, 12/10/61, York Co. Pres. (4/30/62). Transferred To Page's Co.

GUY, ADOLPHUS: Co. A & F. Sgt. Major. Enl. 6/5/61, Yorktown. 4th Cpl., 4/12/62. Sgt. Major, 5/1/62. Transferred To Coke's Co., 7/21/62. Furloughed in Elizabeth City Co., 12/31/61. NFR. Transferred To Dance's Co.

GWINN, WILLIAM D.: Co. F & D. Pvt. Enl. 6/1/61, Williamsburg. In arrest (7/12/62). Present (9/30/62). Transferred To Co. D, 10/6/62.

HALL, SMITH: Co. G. Enl. 5/18/61, Upper Grafton. Pres. Through (4/30/62).

HANSFORD, RICHARD: Co. G. Pvt. Enl. 5/18/61, Upper Grafton. Absent on surgeon's cert., 12/20/61 in York Co. NFR.

HARMAN, JOHN E.: Co. C-2/3. Enl. 5/13/61, Richmond; age 22, roofer. Transferred To C-3. Entered R. E. Lee Camp Soldier's Home, Richmond, 10/14/99, age 66. Dismissed 7/17/07.

HARRARD, WILLIAM T.: Co.C-2/3. Enl. 5/13/61; age 21, blacksmith. Transferred To C-3.

HARRIS, ROBERT: Co. G. Pvt. B. York Co. Enl. 5/18/61, Upper Grafton. Discharged 9/2/62. At G.H. #21, 7/2/62 through 9/2 with typhoid. Also listed as wounded 7/2/62. Age 46, 5'6", dark comp., eyes, grey hair, farmer.

HARTNET, PATRICK: Co. C-2/3. Pvt. Enl. 5/13/61; age 19, laborer. Transferred To C-3.

HARVEY, JAMES A.: Co. Co. C-2/3. Pvt. Enl. 5/13/61, Richmond; age 33, tailor. AWOL (7/15/62). Under arrest in Petersburg (8/31/62). Sent to Ft. Caswell, N.C. under sentence of C.M. to work on fortifications.

HARVEY, MARTIN: Co. C-2. Pvt. Enl. 5/13/61; age 21, laborer. Absent sick in hospital (10/31/61). In arrest in Williamsburg (12/31). Present (2/28/62). AWOL (7/15/62).

HATTON, JOSEPH: Co. C-1. Pvt. Enl. 5/1/61, Manassas. NFR.

HAWES, SAMUEL HORACE: Co. A & K, (Washington Arty & 2nd Richmond Howitzers). B. 6/5/38. Resident of Henrico Co. 2nd lieutenant, Coke's Battery. Pvt and Cpl. 2nd Richmond Howitzers. Assigned as 2nd lieutenant to the Orange Artillery by order of Brigadier General William N. Pendleton, 10/4/62. Detached service to regain absentees till 1/63. Captured at Spotsylvania C.H. 5/12/64; light comp., brown hair, gray eyes, 5'9". . Sent to Belle Plain then to Ft. Delaware 5/17/64. Sent to Hilton Head 8/20/64; to Ft. Pulaski 10/20/64. Back to Hilton Head, 11/19/64 thru 1/19/65. Returned to Ft. Delaware where oath was taken 5/31/65. . Also served in Fry's Orange Artillery and promoted to lieutenant. Married 10/3/67 to Martha Heath. Three children. Married 2) Mary Mayo Blair Fitts. Postwar coal business; director of State Planters Bank; President of Richmond Chamber of Commerce. D. 2/10/22; buried in Hollywood Cem., Richmond.

HAWS, J.: Co. C-2. Pvt. Died 10/17/61, buried in Hollywood Cem., Richmond, VA.

HAZELGROVE, JOHN S.: Co. F. 4th Sgt. Enl. 5/20/61, Williamsburg. Absent on leave (12/31/61). Absent sick in Williamsburg hospital (2/62). Pres. (5/62). Discharged 9/23/62.

HEILMAN, JULIUS F.: Co. C-2. Jr. Ord. Sgt. Enl. 5/13/61; age 30, soldier? Ord. Sgt. (8/31/61). Jr. 2nd Lt. after transferred to C-3, 12/30/62.

HENDERSON, ALEXANDER: Co. C-2/3. Pvt. Enl. 5/13/61; age 25, moulder. Transferred to C-3.

HICKMAN, SAMUEL: Co. A & F. Pvt. Enl. 5/13/61, Hampton. At GH #18, 5/21/62. At GH #21, 5/28/62. Sent to private quarters. Transferred to Coke's Co., 7/21/62. Transferred to Dance's Co.

HIGGINSON, W. MOREHEAD: 2nd Sgt. Co. G.

HILL, CHARLES: Co. K & F. 1st Lt. Son of A. Hill, jr. Enl. 4/21/61, Richmond, 1st Cpl. Elected 1st Lt. from 2nd Co. Richmond Howitzers to Coke's Battery, 5/10/62. Absent sick in Richmond, (9/30/62). 1st Lt. A.A.I.G., District of Texas, New Mexico and Arizona, General J.B. Magruder, commanding. Recommended for major of artillery, 3/31/63. Listed as Major Charles Hill, 8/31/63 by Gen. Magruder. At battle of Galveston rendered signal service. KIA about 11/23/63 at Matagorda Island, Texas. Claim filed 7/15/64.

HILL, CHARLES S.: Co. C-1. 2nd Lt. Enl. 5/1/61, Manassas. Absent detached (11/13/61).

HILLEARY, GEORGE H.: Co. C-1. Pvt. Enl. 5/1/61, Manassas. NFR.

HILLEARY, WASHINGTON M.: Co. C-1. Pvt. Enl. 6/1/61, Manassas. NFR.

HINES, JOHN: Co. F & K. Pvt. Enl. 3/18/62, Richmond. Transferred from Cropper's Co. Transferred to Co. K, 10/6/62. Captured near Gettysburg, 7/5/63.

HOGG, JOHN: Co. G. Pvt. Enl. 5/18/61, Upper Grafton. Absent in Williamsburg hospital, 9/1/61. Furloughed for 30 days, 12/6/61, at home in York Co. Pres. 10/31/62 - (4/30/62). NFR.

HOGG, LEROY: Co. G. Postwar reference.

HOGG, SAMUEL: Co. G. Pvt. Enl. 5/18/61, Upper Grafton. Absent on surgeon's leave 11/20/61 in York Co. With dysentery at Williamsburg hospital 11/22/61 - 12/6/61, furloughed. Suffering from pneumonia 4/4/63. At Chimbo. #3, 4/13/62. Transferred to Danville from Camp Winder 4/22/62. At Danville, 4/22/63. Transferred to Lynchburg 6/26/62. At GH Liberty, Va. 9 & 10/62. NFR.

HOGG, THOMAS: Co. F. Co. F & D. Pvt. Enl. 5/20/61, Williamsburg. Transferred from Coke's Co. To Co. D, 10/6/62.

HOGGE, ADDISON: Co. G. 2nd Cpl. Enl. 5/18/61, Upper Grafton. Reduced to private 4/22/62. Sick w/fever in Williamsburg hospital 4/22/62 - 5/1/62. NFR.

HOLLAND, ALBERT G.: Co. C (Beauregard Rifles). Pvt. B. 1842, Washington, D.C. Enl. 4/23/61 with the Beauregard Rifles at Alexandria. CSR shows enl. 5/1/61, Manassas. Later served w/1st Va. Cav. Mustered out 9/61 and enlisted in the Purcell Battery 2/18/62, at Richmond. Wounded at Mechanicsville. Cpl. 4/63. Transferred to 2nd Maryland (Baltimore Light Artillery), 4/64, by order Secretary of War.

HOLMEAD, CHARLES H.: Co. C-1. Pvt. Enl. 5/1/61, Manassas. NFR.

HOPE, GEORGE WASHINGTON: Co. A. 2nd Lt. Enl. 5/13/61, Hampton. 1st Sgt. Promoted 2nd Lt. to fill promotion of J. Hope to surgeon, 1/7/62. NFR.

HOPE, JESSE P.: Co. A. 2nd Lt. B. 1828, Hampton, Va. Son of Dr. William Hope. Great-grandson of George Hope, Sr. who was sent to the colony of Virginia to see what the feasibility of a colonial navy would be. Later served in the Continental Army and after the Revolution, established the first Masonic Lodge in Hampton. Graduate of U. Va. and Jefferson Medical College, Philadelphia. Doctor of Medicine, 1857. Married in 1857 to Mary Letitia, daughter of Colonel Taylor, of James City County. Seven children from the marriage. Established practice in Hampton, Va. and served as such until the opening of the war. Enl. 5/13/61, Hampton. Furloughed in Richmond, (12/61). Appointed surgeon. Transferred in late 1861 to the medical dept. as field and hospital surgeon. Served most of the war in hospitals in Richmond. After the war returned to Hampton and continued his practice. D. 6/29/83, Hampton, Va.

HORNSBY, BENJAMIN: Co. G. Pvt. Enl. 5/18/61, Upper Grafton. Absent in Williamsburg hospital 9/10/61. D. at home in York Co., 1/20/62 of disease.

HORNSBY, GEORGE: Co. F. Pvt. Enl. 5/20/61, Williamsburg. Deserted 5/6/62.

HORNSBY, JOHN: Co. G. Pvt. Enl. 5/18/61, Upper Grafton. Discharged disabled 9/10/61.

HOWARD, CALTHROP: Co. G. Pvt. B. Elizabeth City Co. Enl. 5/18/61, Upper Grafton. Discharged 7/16/62; dark comp., grey eyes, age 35, 5'11", boatman.

HOWARD, CHARLES W.: Co. A. Pvt. Enl. ? Discharged 1/22/62.
HOWARD, THOMAS C.: Co. G. Pvt. Enl. 5/18/61, Upper Grafton. Absent on surgeon's leave 12/13/61, Williamsburg. Present (2/28/62). NFR.
HUBBARD, JOHN M.: Co. A. Pvt. Enl. 5/13/61, Hampton. Captured at Fox Mill while on a 24 hour pass, 1/18/62.
HUBBARD, LEPRON: Co. A. 6th Cpl. Enl. 5/13/61, Hampton. 4th Cpl. 1/7/62. 3rd Cpl. 4/12/62. Sgt. Major, 1/62. Transferred to Coke's Co. 7/21/62. Pvt. (9/30/62).
HUDGINS, ANTHONY D.: Co. A. Pvt. Enl. 6/5/61, Yorktown. Present through (4/30/62).
HUDGINS, BENEDICT BENJAMIN: Co. G. Pvt. Enl. 5/18/61, Upper Grafton. At soldiers' home in Richmond, 5/30/62. D. 6/4/62.
HUDGINS, WILLIAM T.: Co. G. Pvt. Enl. 6/15/61, Williamsburg. Absent in hospital Williamsburg, 9/20/61 - 10/31/61. Present (12/31/61). At G.H. Danville, 5/15/62 - 9/9/62 as a nurse. NFR. York Co. p.l.
HUDSON, EDWARD I. or ETHELBERT J.: Co. F. Pvt. Enl. 5/3/61, Prince Edward C.H. Absent sick (5/26/62). Elected 1st Lt., 4/62 in Randolph Cav., Co. G, 5th Va. Cav.

INGLEE, JOHN: Co. F. Pvt. Transferred from Cropper's Co. Found as a deserter from DeLourney's Co. And arrested 7/2/62, Richmond. NFR.

JARBOE, GEORGE T.: Co. C-1. Pvt. Enl. 5/1/61, Manassas.
JENKINS, LOTERICK: Co. A. Pvt. Enl. 5/13/61, Hampton. Present through (4/30/62).
JESTER, JOHN Y.: Co. F. Pvt. Enl.6/21/61, Williamsburg. Absent on furlough 12/2 - 1/2/62. Deserter 2/26/65. Oath 2/27/65, Army of the James. Transportation provided from Ft. Monroe.
JOHNSON, ANDREW J.: Co. A. Pvt. Enl. 5/13/61, Hampton. Transferred to Coke's Co., 7/21/62. NFR.
JOHNSON, CORNELIUS: Co. F. 4th Cpl. Enl. 5/20/61, Williamsburg. Absent detailed as blacksmith (5/26/61). 4th Cpl. (10/61). Reduced to ranks, 5/10/62.
JOHNSTON, THOMAS J.: Co. C-2/3. Orderly Sgt. Enl. 5/13/61, Richmond; age 28, merchant. Orderly Sgt., 6/30/61 in place of Howard L. Burrows. Reduced by (8/31). AWOL (7/15/62). Present (8/31/62). Transferred to C-3.
JONES, HENRY F.: Co. G. 4th Sgt. Enl. 5/18/61, Upper Grafton. Present through to transfer to Page's Co.
JONES, HENRY: Co. C-2. Pvt. Enl. 5/13/61, Richmond; age 20. In arrest in Petersburg 12/4/62. Present through transfer to C-3.
JONES, ISAAC L.: Co. A, F & D. 4th Sgt. Enl. 5/13/61, Hampton. 2nd Cpl. (10/31/61). 1st Cpl. (12/31/61). 5th Sgt. 1/7/62. 4th Sgt., 4/12/62. Transferred to Coke's Co. As private, 7/21/62. 1st Cpl. 8/1/62. Transferred to Co. D.
JONES, JAMES, Jr.: Co. C-1. Pvt. Enl. 5/1/61, Manassas. NFR.
JONES, WILLIAM B. "Wildcat": Co. G & F. Captain. Enl. 5/18/61, Upper Grafton. Acting adjutant for Col. Cabell, 4/8/62. Surrendered with General Kirby Smith, 6/7/65. Paroled in Shreveport, La. Resident of York Co.
JONES, WILLIAM S.: Co. A. 6th Sgt. Enl. 5/13/61, Hampton. Absent detailed to Quartermaster Dept., Jamestown Island, 10/17/61. NFR.
JORDAN, HENRY: Co. C-2/3. Pvt. Enl. 7/1/62, Richmond. On sick leave (8/31/62). Deserted 10/15/62. Returned to serve w/C-3.

KALDENBACK, JAMES H.: Co. C-1. Pvt. Enl. 5/1/61, Manassas. NFR.
KELLUM, JAMES: Co. A. Pvt. Enl. 5/13/61. NFR.
KELLY, GEORGE WASHINGTON: Co. C-2 & H. Cpl. Enl. 3/10/62, Fincastle. Transferred from Ewing's Co. Pvt. (8/31/62). Transferred to Co. H, 10/28/62.
KELLY, S. F.: Co. C-2 & H. Pvt. Enl. 3/10/62, Fincastle. Transferred to Co. H, 10/28/62.
KENDALL, JOHN W.: Co. A. 1st Sgt. Enl. 5/13/61, Hampton. 5th Sgt. to 2nd Sgt., 10/31/61. 1st Sgt. 1/7/62. At G.H., Camp Winder, 5/7 - 6/8/62. NFR.
KING, MILES H.: Co. A. Pvt. Enl. 6/28/61, Yorktown. Present through (4/62).

LACY, WILLIAM D.: Co. C-2/3. Pvt. Enl. 5/13/61, Richmond; age 18. Transferred to Co. C-3.
LANCASTER, SAMUEL G.: Co. C-1. Pvt. Enl. 5/1/61, Manassas. Discharged 9/10/61. NFR.
LATIMER, SAMUEL: Co. A. Pvt. Enl. 5/13/61, Hampton. Detailed to QM Dept. 6/1/61, Yorktown. At Chimbo. #2, Richmond, 12/5/64 with pneumonia. NFR.
LAWSON, ALEXANDER G.: Co. F & K. Pvt. Enl. 5/20/61, Williamsburg. Transferred to Co. K, 10/6/62. NFR.
LAWSON, DENNIS: Co. A. Pvt. Enl. 9/3/61, Ship Point. Present through (4/30/62).
LAWSON, JAMES: Co. A. Pvt. Enl. 9/1/61, Ship Point. At G.H. Danville w/diarrhea, 5/15 - 6/20/62. NFR.
LAWSON, JAMES S.: Co F & G. 1st Lt. Enl. 5/20/61, Williamsburg. Absent sick and in arrest in hospital (12/31/61). Entire record is tainted w/uncertainties. Apparently originally student of U.S. Naval

Academy. Also claimed he graduated from V.M.I., 1853. Later Lt., U.S.N. Sent to take charge of heavy batteries at Washington, D.C. There until Burnside expedition when he resigned, which in fact conflicts with his enlistment date by almost a year. Crossed the Potomac below Romney There is a slip in his record that states he was a 1st Lt., 4th Regt. Art. Of the 9th Brigade, 4th Div. Va. Militia, 5/11/61. In an incident on 12/13/61, Lawson had been sick in Williamsburg for a few days. When he returned to camp to check in, Captain Garrett ordered him to remain in camp. He refused and the captain demanded he surrender his sword and consider himself placed under arrest. Called upon Garrett to "take off his straps" and he would fight him over the matter. Lawson claimed that he would surrender his sword only to General McLaws or Col. Cabell. Sgt. Josiah N. Wilson and Sgt. Palmer witnessed the incident. Charged with 1) disrespect of commanding officer; 2) disobedience; 3) conduct prejudice to good order and discipline, 12/19/61. On 1/17/62, after a general court martial, Lawson was found guilty on all counts and ordered dismissed from the service. On 3/18/62, Lawson supposibly obtained permission slips to go to New Orleans to recruit for the company for 60 days. Noted in CSR as a recruiting officer, 4/11/62. At New Orleans, La., 3/30/62, wrote letter to the Secretary of War of his intent to raise a battalion of 500 from Texas. Arrested as a spy while in Goliad, Victoria Co.?, Texas, 5/62. In Houston jail 5/14/62 as Federal spy. NFR.

LAWSON, JOHN W.: Co. A & F. 1st Sgt. B. 9/13/37, James City Co., the son of James S. and Sallie (Hankins) Lawson. Attended College of William and Mary, the University of Virginia, and the University of New York, graduating in March, 1861. Enl. In Washington Artillery, 5/20/61, Williamsburg upon returning home. After May, 1862, he was transferred to the medical dept. as asst. surgeon and promotion to surgeon followed a few months later, with the rank of major. Assigned to the 12th N.C. regiment. Gave up post of orderly sgt. With the battery on 5/10/62 for commission he received 6/12/62. Following the war began a practice in Isle of Wight County before retiring to his farm, "The Rocks," upon the James River. Resided in Smithfield since 1893. Served as a delegate from the county for four continuous sessions and one term as senator from the district by election in 1873, and in 1883, was reelected to the house of delegates. Married 1/30/77 to Margaret Norfleet Urquhart, of Southampton Co. Six children. In 1890, he was elected to Congress from the second district, serving one term. Twice served on board of visitors of the agricultural and mechanical college of Virginia, and for many years also with VMI and College of William and Mary. President of the board of visitors at College of William and Mary. President of the Smithfield alliance company.

LAWSON, LEVIN: Co. A. Pvt. Enl. 5/13/61, Hampton. Present through (4/30/62). NFR.

LAWSON, THOMAS: Co. F. Pvt. Enl. 5/20/61, Williamsburg. Absent sick (7/12/62), Richmond. NFR.

LAWSON, WILLIAM: Co. A. Pvt. Enl. 9/1/61, Sewell's Point. Present through (4/30/62). NFR.

LEATHERS, JOHN W.: Co. C-2. Pvt. B. Orange Co. Enl. 5/13/61, Richmond; age 18, cooper. Absent sick (7/15/62) at Camp Winder. Discharged 8/5/62; age 17, 5'5", dark comp. and hair, blue eyes, farmer.

LESTER, JOHN Y.: Co. F. Postwar reference.

LEWIS, ROBERT ESTON: Co. C-2. 5th Sgt. Res. of Sussex Co. B. in Goochland Co., 2/1/1825. Henrico County clerk at outbreak of the war. Enl. 5/13/61, Richmond; age 36, clerk. 2nd Cpl. (8/31/61). 5th Sgt. (10/31/61). Pvt. (7/15/62). Discharged 8/13/62; age 38, 5'10", dark comp. and hair, black eyes, clerk. D. 1876 in Richmond.

LINDENBURG, LEWIS: Co. F. Pvt. Enl. 3/18/62, Richmond. Transferred from Cropper's Co. Deserted 8/20/62.

LINDSAY, THOMAS H.: 6th Cpl. Co. F & D. Enl. 5/20/61, Williamsburg. Transferred to Co. D, 10/6/62.

LINDSAY, WILLIAM: 1st Cpl. Co. F. B. Williamsburg. Enl. 5/20/61, Williamsburg; 2nd Cpl. 1st Cpl. (7/12/62). At G. H. #2 w/hepatitis, 6/2 - 7/19/62. To private quarters. Discharged 7/22/62; age 29, 5'10", dark comp., hair and eyes, mechanic.

LINKENHOGER, J. W.: Co. C-2 & H. Enl. 3/10/62, Fincastle. Transferred from Ewing's Co. Transferred to Co. H., 10/28/62.

LIPSCOMB, JAMES H.: Co. C-2 & H. Enl. 5/13/61, Richmond; age 19, carpenter. Transferred to Wyatt's Co., 10/28/62.

LIPSCOMB, SPOTSWOOD: Co. F, K, & E. 4th Sgt. Enl. 5/20/61, Williamsburg. Sick in hospital (12/31). Pres. (5/26/62). Cpl. 3/10/62. 4th Sgt. 5/10/62. Transferred to Co. K, 10/6/62. WIA Fredericksburg, 12/13/62.

LITTLE, CHARLES M.: Co. C-2 & H. Pvt. Enl. 3/10/62, Fincastle. Transferred from Ewing's Co. Transferred to Co. H, 10/28/62.

LONG, MELCHOIR M.: Co. C-1. 5th Sgt. Enl. 5/1/61, Manassas. 5th Sgt. , 9/1/61. Detailed 10/1/61. Discharged 10/1/61. Another slip states discharged 1/17/62. NFR.

LOTTERICK, JENKINS: Co. A. Postwar reference.

LOWE, DANIEL C.: Co. C-1. Pvt. Enl. 5/1/61, Manassas.

LYNCH, EDWARD: Co. A. Pvt. Enl. 5/13/61, Hampton. Present through (4/30/62). NFR.

LYONS, JERRY: Co. C-2. Pvt. B. Ireland. Enl. 5/13/61; age 22, laborer. Discharged 7/18/62; light comp. and hair, blue eyes.

MACKALL, ROBERT: Co. C-1. Pvt. Enl. 5/1/61, Manassas. Discharged 10/3/61. At G.H. Orange C.H., 10/9 - 10/10/61. Examined 5/3/62, found capable again for service. See 1st Stuart Horse Artillery. Color Bearer. Supposibly assisted Lt. Colonel John Pelham after he was mortally wounded at Kelly's Ford, Va. Resided after war in Culpeper Co., Va.

MADISON, JAMES: Co. C-2/3. Pvt. Enl. 5/13/61, Richmond; age 24, carpenter. Transferred to C-3.

MAGRUDER, GEORGE: Co. C-1. 3rd Cpl. Enl. 5/1/61, Manassas. Sick in Richmond (11/13/61). NFR.

MAHONE, ABEL S.: Co. F & K. Pvt. Enl. 5/20/61, Williamsburg. Transferred to Co. K, 10/6/62.

MAHONE, JAMES THOMAS: Co. F. Pvt. Enl. 6/1/61, Williamsburg. Absent sick in hospital (10/31/61). In Seminary hospital, Williamsburg w/fever, 11/20 - 12/1/61. Deserter. NFR.

MALONEY, PATRICK: Co. C-2/3. Pvt. Enl. 5/13/61; age 35, laborer. Pres. In arrest (7/15/62). Absent under arrest (8/62). Present (10/31/62). Transferred to C-3.

MARRON, JOHN C.: Co. C-1. Pvt. Enl. 6/13/61, Manassas. Detached 10/1/61 per S.O. #419.

MARROW, DANIEL G.: Co. A. Pvt. Enl. 5/13/61, Hampton. Furloughed in Richmond, 12/24/61. NFR.

MARSHALL, JOHN P.: Co. C-1. Pvt. Enl. 6/19/61, Manassas. Absent sick in Charlottesville hospital (11/13/61). NFR.

MARSHALL, ALPHEUS: Co. C-2/3. Pvt. B. in Richmond. Enl. 5/13/61, Richmond; age 28, coach maker. Absent in arrest (8/31/62). Pres. (12/31/62). Transferred to C-3. 6' ½", light comp., blue eyes, dark hair.

MARTIN, JAMES: Co. A. Pvt. Enl. ? Pay record 7/64. NFR.

MARTIN, ENOCH TIMOTHY: Co. C-2, Enl. 5/61 in James City Co., VA. B. 1840 in James City Co., VA. D. 1920.

MATTHEWS, HENRY H: Co. C-1. Pvt. Enl. 5/1/61, Manassas. Discharged 10/22/61.

MAYNARD, JAMES E.: Co. G. Pvt. Enl. 6/21/61, Williamsburg. Present through 3/27/62 at Seminary hospital Williamsburg. Returned 4/24/62. Pres. through (4/30/62). D. 1862. York Co. p.l.

McCANDISH, THOMAS P.: Co. G. Postwar reference.

McCORMICK, HENRY A.: Co. C-1. Pvt. Enl. 5/1/61, Manassas. NFR.

McDEARMAN, BERNARD: Co. C-2/3. Pvt. Enl. 5/13/61, Richmond; age 26, shoemaker. AWOL (7/15/62). Deserted 4/1/62.

McHENRY, JAMES, Jr.: Co. C-1. Pvt. Enl. 5/1/61, Manassas. NFR.

McINTOSH, ROBERT: Co. G. Pvt. Enl. 7/1/61, Williamsburg. Absent 9/15/61 - 10/31/61. Transferred to Captain Lee, served in Co. Until 9/15/62 and was absent at home sick on leave.

MERGHAN, ROGER: Co. C-2/3. Pvt. Enl. 5/13/61, Richmond; age 24, laborer. Deserted 3/15/62. AWOL (7/15/62). In arrest in Richmond, (10/31/62).

MERRITT, LEWIS T.: Co. C-2/3. Pvt. Enl. 3/10/62, Fincastle. Transferred from Ewing's Co. Transferred to Co. H, 10/28/62.

MILLER, CHARLES: Pvt. Co. A, 1st VA Lt. Art. Paroled on 6/2/65, described as 5'7-", dark hair, dark eyes, light comp.

MILLS, JOHN: Co. F & K. Pvt. Enl. 3/18/62, Richmond. Transferred from Cropper's Co. Transferred to Coke's Co., 6/11/62. Transferred to Co. K, 10/6/62.

MINTER, JAMES J.: Co. C-2/3. Pvt. Enl. 5/13/61, Richmond; age 21, teamster. Transferred to C-3.

MOORE, MERRITT S.: Co. G. Pvt. Enl. 7/1/61, Williamsburg. Present through 10/31/61. NFR.

MOORE, RICHARD A.: Co. F. 2nd Sgt. Enl. 5/20/61, Williamsburg. 4th Sgt. (12/31/61). 2nd Sgt. (7/12/62). Transferred to Co. K, 10/6/62. Wounded (flesh) in the knee, 12/13/62, Fredericksburg. Died in hospital 5/29/63 of typhoid.

MORELAND, HIGGINSON W.: Co. G. 2nd Sgt. B. York Co. Enl. 5/18/61, Upper Grafton. Absent w/leave for 4 days in York Co., 12/30/61. Dark comp., black eyes and hair, age 35, 5'10", farmer.

MORGAN, JOHN S.: Co. G. Pvt. Enl. 6/1/61, Williamsburg. Absent sick at home on surgeon's leave, 9/10 - 10/31/61. D. At home, 12/26/61 of disease.

MORRISON, JOHN H.: Co. G. 1st Sgt. B. Williamsburg. Enl. 5/15/61, Upper Grafton. At GH #21, 6/10, chronic diarrhea. Discharged 8/13/61; age 40, 5'9", dark comp., hair and eyes, merchant.

MORROW, DANIEL GARROW: Co. A. B. 1/13/1836 D. 7/7/1914, buried in the St. John's Cem., Hampton, VA.

MOULTRY, JAMES: Co. F & D. Pvt. Enl. 3/18/62, Richmond (Cropper's Co). Transferred to Co. F. Transferred to Co. D, 10/6/62. Sent to hospital at Gordonsville, 11/27/62.

MURPHY, DANIEL: Co. C-2. Pvt. B. in Ireland. Enl. 5/13/61, Richmond; age 36, laborer. Discharged 8/13/61; age 43, 5'8", dark comp., grey eyes, light hair.

MURRAY, EDWARD C.: Co. C-1. Pvt. Enl. 5/1/61, Manassas. Substitute for Carl Yulee who was discharged 10/1/61. Pres. (11/13/61).

MYERS, CHARLES F.: Co. C-1. Pvt. Enl. 6/13/61, Manassas. At soldiers home in Richmond, 7/26/62 Sent to hospital in Danville, 7/29/62. NFR.

NORRIS, WALTER, A.: Co. C-1. Enl. 5/1/61, Manassas. Absent sick in Richmond, (11/13/61). Deserter at Orange C.H., 10/9/61 - 10/10/61. Then to general hospital. NFR.

OLIVER, JOHN J.: Co. C-2/3. Pvt. Enl. 6/1/61, Baptist College near Richmond; farmer. Absent sick with leave (7/15/62). Cpl. 8/10/62. Transferred to C-3.

OLIVER, J. W.: Co. C-2 & H. Pvt. Enl. 3/10/62, Fincastle. Transferred from Ewing's Co. Transferred to Co. H, 10/28/62.

O'NEAL, HIRAM O.: Co. C-1. Pvt. Enl. 5/1/61, Manassas. Absent sick in Richmond, (10/61). Present (11/13/61). NFR.

O'NEIL, SIMON: Co. C-2/3. Pvt. Enl. 5/13/61, Richmond; age 18, laborer. AWOL (2/28/62). In arrest in Richmond (7/15/62). Transferred to C-3.

OTTO, JOHN: Co. F, K, & B. Pvt. Enl. 3/18/62, Richmond, Cropper's Co. Transferred to Williamsburg Art., 6/11/62. Transferred to Co. K, 10/6/62. See also 24th Battalion Virginia Partisan Rangers.

OTEY, GEORGE W.: Co. F. Pvt. Enl. 6/22/61, Williamsburg. Absent sick in hospital (10/31/61). Discharged 2/18/62.

PALMER, HENRY E.: Co. F. Lt. Enl. 5/22/61, Williamsburg. 5th Sgt. (10/61). Elected Lt. of artillery (5/26/62).

PALMER, WILLIAM T.: Co. F & K. Pvt. Enl. 5/20/61, Williamsburg. Transferred to Co. K, 10/6/62.

PARKER, GEORGE T.: Co. A, F & D. Pvt. Enl. 5/13/61, Hampton. Transferred to Coke's Co., 7/21. Transferred to Co. D, 10/6/62.

PARKER, HENRY E.: Co. A. Postwar reference.

PARKER, JOSEPH: Co. C-1. Pvt. Enl. 5/1/61, Manassas. Present through (11/13/61). NFR.

PARRY, SAMUEL A.: Co. F. Pvt. Enl. 6/21/61, Williamsburg. Absent sick in Williamsburg (10/31/61). At Seminary Hospital in Williamsburg, 11/20 - 11/29/61. NFR.

PEEK, CHARLES SMITH: Co. A. Pvt. Son of Thomas and Janet Hope Peek. Enl. 6/25/61, Yorktown. Absent detailed to Mssr. Talbot and Brother in Richmond, 12/10/61. Later Asst. Engineer in C.S. Navy. After the war was a machinist and later engaged in mercantile pursuits. D. 1887.

PEEK, EDWARD K.: Co. A. 3rd Sgt. Son of Thomas and Eliza Peek. Enl. 5/13/61, Hampton; age 28; occupation mechanic. 1st Cpl. (10/31/61). 5th Sgt. 12/1/61. 3rd Sgt. (2/28/62). 9/1/62 enl. in Co. I, 32nd Va. Inf. Wounded at Sharpsburg (slightly). 1st Sgt. 9/10/63. Recommended to promotion as 2nd lieutenant 5/13/63, but not promoted. Paroled at Appomattox C.H. 4/9/65. Resident of the Soldier's Home in Richmond, Va., 1894. Died 6/7/1900. Buried in Hollywood Cem.

PEEK, JOHN L.: Co. A. 2nd Sgt. Son of Thomas and Eliza Kirby Peek. Enl. 5/13/61, Hampton. Discharged 9/20/61. After serving with the Washington Artillery, attached to War Dept. Died, 1890.

PENDLETON, SAMUEL H.: Co. K & F. 1st Lt. Enl. 4/21/61, Richmond in the Richmond Howitzers. Elected 1st Lt. Coke's Co., 5/10/62. Pres. (9/20/62).

PETTITT, WILLIAM F.: Co. G. 5th Sgt. Enl. 5/18/61, Upper Grafton. With fever in Seminary Hospital, Williamsburg, 11/29/61 - 12/10/61. Absent on surgeon's leave in York Co., 4/22/62.

PHILLIPS, JACOB: Co. F & D. Pvt. Enl. 3/18/62, Richmond, Cropper's Co. Transferred to Coke's Co., 6/11/62. Transferred to Co. D, 10/6/62.

POLLARD, GEORGE B.: Co. C-2/3. 4th Sgt. Enl.5/13/61, Richmond; age 19. Reduced at own request, 6/30/61. 4th Sgt. (7/15/62. Transferred to C-3.

POLLARD, LEONIDAS C.: Co. C-2 & H. Pvt. Enl. 5/13/61, Richmond; age 18, tinner. Transferred to Co. H, 10/28/62.

POTTS, JOHN: Co. F & K. 3rd Sgt. Enl. 5/20/61, Williamsburg. Cpl. to 3rd Sgt., 5/30/62. Transferred to Co. K, 10/6/62.

PRESSON, WILLIAM M.: Co. G. Cpl. B. York Co. Enl. 6/5/61, Williamsburg. Present through 3/27/62 when Cpl. Discharged 7/15/62, York Co.; age 36, 5'10", dark comp. and hair, black eyes, merchant. York Co. p.l.

QUARLES, GEORGE WASHINGTON: Co. C-2/3. 4th Cpl. Enl. 5/13/61, Richmond; age 23, machinist. AWOL (7/15/62). Pres. (10/31/62). Furloughed for 15 days, 12/30/62. 4th Cpl. (12/31/62). Transferred to C-3.

RANDOLPH, CHARLES H.: Co. C-1. Pvt. Enl. 7/1/61, Camp Stevens. Detached 10/1/61. D. about 10/61.

RATCLIFFE, WALTER H.: Co. C-1. Pvt. Enl. 5/30/61, Manassas. Commissioned captain, P.A.C.S.

REGNAULT, CHARLES H.: Co. C-2. Cpl. Enl. 5/13/61, Richmond; age 39. Cpl. Transferred to C-3.

REID, WILLIAM H.: Co. A. Pvt. Enl. 5/13/61, Hampton. Present through (4/30/62).

RICHARDSON, JOHN W.: Co. A. Pvt. Enl. 5/13/61, Hampton. Present through (4/30/62).

RICHARDSON, NATHAN: Co. F. 2nd Lt. Enl. 5/20/61, Williamsburg. Present through (5/26/62).

RICHARDSON, R. H.: Co. A. Postwar reference.

RICHEY, JAMES J.: Co. C-1. Pvt. Enl. 5/28/61, Manassas.

RIGGINS, THOMAS P.: Co. G. Pvt. Enl. 6/15/61, Williamsburg. Sick in Williamsburg hospital, 4/13/62. At GH #21, Richmond with diarrhea. Transferred to Scottsville, 7/10. D. 8/16/62, Syphilis, Scottsville, Va.

RIND, SEALBROOK, Jr.: Co. C-1. Pvt. Enl. 6/21/61, Manassas.

RIPLEY, LEMUEL: Co. A. Postwar reference.

RITTER, WILLIAM BAILEY: Co. I, C-2, & E. Captain. Enl. 4/25/61, Richmond; age 24, lawyer. 2nd Lt. on Sand's rolls, 5/13/61. Elected captain, 4/30/62. Ordnance Officer, Daniels Brigade, Rodes'' Division, 2nd Corps, 3/18/64. Relieved 7/14/64. POW at Charlotte, N.C., 5/13/65.

ROACH, CHRISTOPHER: Co. F & E. 1st Cpl. Enl. 5/20/61, Williamsburg; 1st Cpl. On leave 12/31 - 1/2/62. Discharged 8/21/62 or 9/13/62. Entered R.E. Lee Camp Soldiers Home, Richmond, 4/10/85, age 68. Sent to Williamsburg, 6/19/84.

ROBERTSON, WALTER H.: Co. I & C-2. 1st Lt. Enl. 4/25/61, Richmond; age 20, lawyer. 2nd Lt. in Sands' Co., 5/13/61. 1st Lt. 4/30/62 in place of W.W. Fontaine. Signs commanding company, 98/31/62). On leave for 12 days, 12/2 - 1/2/62. Ord. Officer, Pettigrew's brigade, 8/31/63. On leave 8/31/63, 30 days. Leave extended 30 days, 10/28/63. In home recovering from wounds received at Gettysburg, 10/30/63. Returned to brigade, 12/4/63. On Gen. Kirkland's staff on leave extended for 25 days, 12/24/63. Ordered to Demopolis, Alabama to general Polk for duty with General S.G. French, 3/15/64. Resigned about 10/15/64. NFR.

ROBERTSON, WILLIAM H.: Co. F. 2nd Cpl. Enl. 4/22/61, Pittsylvania Co. 2nd Cpl. 5/20/62. Present through (9/30/62).

ROGERS, JOHN WALTER: Co. F & E. Pvt. Enl. 5/20/61, Williamsburg. Discharged 12/12/61.

ROWAN, JOHN: Co. C-2/3. Pvt. Enl. 6/1/61, Baptist College. On sick leave (8/31/62). Transferred to Co. H, 10/28/62.

SANDS, JOHNSON H.: Co. C-2. Captain. Enl. 5/13/61. Listed as commissioner recommended as drill master and or captain of artillery by G.W. Randolph and J.B. Magruder, 2/63. On list of prisoners from various organizations previously confined in E.D.M. prison who volunteered for Wise Legion for the defense of Richmond against Sheridan's raid, 1864. Were pardoned by President Jefferson Davis, 8/3/64. D. after 1908.

SANFER, JAMES: Co. F. Pvt. Enl. 6/20/61, Williamsburg. Absent sick (7/12/62). Present (9/30/62). At GH #21, 6/14 - 6/28, Danville. Also from 6/29 - 7/13/62. NFR.

SAVAGE, EDMUND E.: Co. A. Pvt. B. 3/1/1817. Enl. 5/15/61, Hampton. On 20 day furlough, 1/18/62. D. 7/2/1909, buried in St. John's Cem., Hampton, VA. Awarded Confederate Cross of Honor on 5/27/1905.

SCHAEFFER, FRANK B.: Captain. Enl. 5/1/61, Manassas. Detached 9 - 11/13/61.

SCHAWNESEY, JOHN D.: Co. G. Pvt. B. England. Enl. 6/5/61, Williamsburg. Detailed as blacksmith at Young's Mill, 1/30/62. Captured in Page County, Va., 7/13/62; age 43, 5'10", dark comp., grey eyes, brown hair, blacksmith.

SCHREAVES or SCHAVIS, GEORGE: Co. G & F. Pvt. Enl. 5/18/61, upper Grafton. Absent on surgeons leave, 10/31/61, Williamsburg. Present (4/62).

SEIMMS, FREDERICK D.: Co. C-2. Pvt. B. England. Enl. 5/13/61, Richmond; age 18, oculist. Discharged 7/20/62; age 17, 5'6", fair comp., grey eyes, light hair, artificer.

SEMPLE, E. A.: Co. A. Postwar reference. B. 1812. D. 1883, buried in the St. John's Cem., Hampton, VA.

SEMPLE, GEORGE WILLIAM: Co. A. 1st Lt. B. 1812 D. 1883. Buried in St. John's Cem., Hampton, VA.

SHACKLEFORD, LAWSON: Co. C-2/3. Cpl. Enl. 5/13/61; age 39, machinist. Cpl. 11/1/61. Transferred to C-3. KIA Cold Harbor, 1864.

SHEILD, WILLIAM H.: Co. G. Postwar reference.

SHELTON, CHARLES C.: Co. A. Pvt. Enl. 5/13/61, Hampton. Detailed 12/61. Discharged and joined C.S. Navy per S.O.# 89, 4/18/62.

SHERMAN, EDWARD A.: Co. C-2/3. B. New Kent Co. Enl. 5/13/61, Richmond; age 37, carpenter. Transferred to C-3. D. 7/29/64 of wounds received 7/28; age 38, 5'5, dark comp., eyes, hair.

SHERWELL, WILLIAM: Co. G. 4th Cpl. B. Nashville, TN. Enl. 5/18/61, Upper Grafton. At GH #21, 5/9 - 5/13/62 with fever. Discharged 7/15/62; age 17, 5'8", light comp. and hair, blue eyes, student.

SHIELD, ORLANDO F.: Co. G & F. 2nd Lt. Enl. 5/18/61, Upper Grafton. Absent sick in Williamsburg hospital, 4/24/62. D. in Williamsburg of typhoid fever. Claim filed by father, William H. Shield, $459.00 due.

SHIELD, WILLIAM H.: Co. G & F. Pvt. Enl. 6/25/61, Williamsburg. Detached as apothecary in Williamsburg, 8/31/61 - 4/62. NFR.

SMITH, CHARLES L.: Co. A. Captain. B.1818. Enl. 5/13/61, Hampton. D. 1876. Buried in the St. John's Cem., Hampton, VA.

SMITH, GEORGE W.: Co. G. Pvt. Enl. 5/18/61, upper Grafton. Present through (4/30/62).

SMITH, JOHN W.: Co. C-2/3. Pvt. Enl. 5/13/61, Richmond; age 24, teamster. Transferred to C-3.

SMITH, JOSEPH: Co. C-2/3. Pvt. Enl. 5/13/61, Richmond. Present in arrest, (7/15/62). $6.00 retained as fine from CM (8/31/62). NFR.

SMITH, THOMAS F.: Co. G. Pvt. B. in Matthews Co. Enl. 6/18/61, Williamsburg. Absent sick at home in York Co., 12/20/61. Present (4/30/62). Absent on surgeon's cert., 5/10/62. Discharged 7/15/62; age 37, 5' 8 1/4", dark comp. and eyes, brown hair, ship carpenter. York Co. p.l. D. 1896.

SMITH, WESLEY: Co. G. Pvt. B. in Matthews Co. Enl. 5/18/61, Upper Grafton. Age 36, 5'10", dark comp., grey eyes, light hair, boatman.

SMITH, WILLIAM: Co. G. Pvt. Enl. 5/18/61, Upper Grafton. Absent sick at home on surgeon's leave 9/16 - 12/61. Present (2/28/62). NFR.

SMITHERS, WILLIAM G.: Co. A. Pvt. Enl. 5/13/61, Hampton. NFR.

SMOOT, JOSEPH K.: Co. C-1. Cpl. Enl. 5/1/61, Manassas. NFR.

SMYTH, HENRY B.: Co. C-2. Sgt. B. Wytheville. Enl. 5/13/61, Richmond; age 35, 5'10", dark comp. and hair, green eyes, civil engineer.

SPENCER, LEWELLYN: Co. F. Pvt. Enl. 6/18/61, Williamsburg. Absent sick at home (10/31/61). Discharged 12/61.

SPICER, JAMES F.: Co. C-2 & H. Pvt. Enl. 5/13/61; age 18, finisher. Transferred to Co. H, 10/28/62.

STARKIE, GEORGE WASHINGTON: Co. C-2/3. Pvt. Enl. 5/13/61, Richmond; age 28, engineer. AWOL (7/15/62). Present (8/31/62). Transferred to C-3.

STERLING, WILLIAM: Co. G & F. Pvt. Enl. 6/1/61, Williamsburg. Absent at home in Williamsburg 10/21 - 10/31/61. Discharged 11/19/61, Young's Mill.

STOKES, HEZEKIAH: Co. G. Pvt. B. Gloucester Co. Enl. 5/18/61, Upper Grafton. Discharged 7/15/62; age 39, 5'8", dark comp. and hair, grey eyes, farmer.

STROUBE, JOHN: Co. F. Pvt. Enl. 5/22/61, Williamsburg. Present through (10/31/61). NFR.

STROUD, WILLIAM: Co. G. Pvt. B. York Co. Enl. 5/18/61, Upper Grafton. Absent at home in York Co. With surgeon's leave, 2/27/62. Present (4/62). Transferred to Lynchburg with typhoid, 7/7/62. Discharged 8/18/62; age 43, 5'9", dark comp., hair, and eyes, farmer.

SULLIVAN, MICHAEL: Co. C-2/3. Pvt. Enl. 5/13/61; age 21. Laborer. AWOL (7/15/62). Deserted 3/15/62 NFR.

SULLIVAN, TIMOTHY: Co. C-2. Pvt. B. Cork Co., Ireland. Enl. 6/1/61, Baptist College; laborer. Discharged 8/13/62; age 48, 5'5", dark comp., grey eyes, brown hair, farmer.

SWARTZ, GEORGE: Co. G. Pvt. Enl. 5/18/61, Upper Grafton. NFR.

SYKES, WILLIAM: Co. A. Pvt. Enl. 5/13/61, Hampton. Present through (4/30/62). NFR.

TABB, THOMAS: Co. A. 2nd Lt. Enl. 5/13/61, Hampton, 3rd Lt. 2nd Lt. (2/28/62). Present through (4/30/62).

TAYLOR, JAMES: Co. C-2. Pvt. Enl. 5/13/61; age 27, carpenter. Discharged 8/13/61. NFR.

TAYLOR, NATHANIEL, Jr.: Co. G. 3rd Sgt. B. in Yorktown. Enl. 5/18/61, Upper Grafton. Discharged 7/15/62; age 35, 6', dark comp. and hair, grey eyes, farmer.

TAYLOR, REUBEN: Co. F. Pvt. Enl. 9/23/61, Williamsburg. Sick in hospital (12/31/61). Present (5/62). Suffering from carbuncle 12/14 - 1/3/62 in Williamsburg hospital.

THOMAS, GEORGE J.: Co. C-1. Pvt. Enl. 6/11/61. Manassas. Discharged 9/18/61. NFR.

THOMAS, NOAH: Co. A. Pvt. Enl. 5/13/61, Hampton. Pres. through (2/28/62). NFR.

THOMAS, WILLIAM: Co. G. Pvt. Enl. 5/18/61, Upper Grafton. Present through (4/30/62).

TINNEY, JOHN: Co. F. Pvt. Enl. 5/20/61, Williamsburg. Transferred to Co. K, 10/6/62.

TOLSON, ALBERT: Co. C-1. Pvt. Enl. 5/1/61, Manassas. NFR.

TOPPING, JAMES N.: Co. G. Enl. 5/18/61, Upper Grafton. Pres. through (4/30/62). NFR. York Co. p.l.

TRAX, ALEXANDER: Co. F. Pvt. Enl. 3/18/62, Richmond in Cropper's Co. Transferred to Co. F., 6/11/62. Deserted 8/20/62.

TURLINGTON, WILLIAM G. or T.: Co. A & F. Pvt. Enl. 5/13/61, Hampton. D. in hospital, 7/23/62. Furloughed in Elizabeth City Co., 12/28/62.

VAN PELT, ROBERT: Co. A. Postwar reference. Buried in the Elk View Cem., Clarksburg, WV.

VANDEVENTER, JAMES: Co. F & D. 4th Cpl. Enl. 5/20/61, Williamsburg. Sick in hospital (12/61). Pres. (5/62). 4th Cpl. 5/10/62. Transferred to Co. D, 10/6/62.

VANNAME, PETER M.: Co. F. QM Sgt. Enl. 5/20/61, Williamsburg. 5th Sgt. 3/10/62. QM Sgt. (7/62). Detailed 5/10/62. Transferred to Co. K, 10/6/62.

WALLS, ELIJAH B.: Co. F. Pvt. Enl. 6/22/61, Williamsburg. Sick in hospital (12/61). NFR.

WALLS, LITTLEBERRY: Co. F. Pvt. Enl. 6/1/61, Williamsburg. Pres. through (9/62). NFR.

WALTRIP, BENJAMIN: Co. F. Pvt. Enl. 5/20/61, Williamsburg. Pres. through (9/62).

WALTRIP, ROBERT: Co. F. Pvt. Enl. 5/6/61, Burnt Ordinary, on retreat from Williamsburg. Present through (9/62).

WARD, J. W.: Co. A. NFR.

WELSH, THOMAS: Co. C-2/3. Pvt. Enl. 5/13/61, Richmond; age 24, laborer. AWOL (7/15/62). Deserter 4/1/62. NFR.

WESTHEIMER, HARRIS: Co. A. 2nd Cpl. Enl. 5/13/61, Hampton. 4th Cpl. (12/61). 3rd Cpl. (2/62). 2nd Cpl. (4/62). NFR.

WESTWOOD, WILLIAM T.: Co. A. Pvt. Enl. 5/13/61, Hampton; age 25. Transferred to Carter's Co. Wounded at Somerville Ford, 9/14/63. Invalid Corps, 5/4/64. Detailed to clerical duty in QM dept., 6/20/64.

WILDT, CHARLES: Co. C-2 & H. Enl. 5/13/61, Richmond; age 38, cabinet maker. AWOL (7/62). Sick leave (8/62). Transferred to Co. H, 10/28/62.

WILLIAMS, HENRY A.: Co. C-1. Pvt. Enl. 5/1/61, Manassas. NFR.

WILLIAMS, PETER D.: Co. F. Pvt. Enl. 5/20/61, Williamsburg. AWOL (5/26/62). Deserted 5/6/62.

WILLIS, BAILEY: Co. A. Pvt. Enl. 9/3/61, Sewell's Point. Present through 4/62. NFR.

WILSON, CHARLES, Jr.: Co. C-1. Pvt. Enl. 6/16/61, Manassas. Discharged 10/1/61. NFR.

WILSON, JAMES: Co. F. Enl. 5/20/61, Williamsburg. D. 1862. Coffin and hearse provided, 12/6/62.

WILSON, JOSIAH N.: Co. F & K. 1st Sgt. Enl. 5/20/61, Williamsburg. 2nd Sgt. Absent on leave (12/62). 1st Sgt. 5/10/62. In Surry Co., sick, 8/24/62. Transferred to Co. K, 10/6/62.

WILSON, THOMAS: Co. F. Pvt. B. James City Co. Enl. 5/20/61, Williamsburg. Absent sick, (10/61). Discharged 12/61; age 22, 5'10", shallow comp., dark hair and eyes, farmer.

WILSON, WILLIAM S. L.: Co. A. Postwar reference.

WILSON, WILLIAM: Co. F. Enl. 5/20/61, Williamsburg. Present through (9/30/62). NFR.

WINEBERGER, CYRUS H.: Co. A. Pvt. Enl. 5/13/61, Hampton. Detached as baker "in a private concern." (T.S. Armistead, Williamsburg.)

WINEBERGER, WILLIAM W.: Co. A. Pvt. Enl. 5/13/61, Hampton. Present through (4/30/62).

WOOD, EDWARD C.: 3rd Sgt. Co. A. B. 1838. Enl. 5/13/61, Hampton. Present through (4/30/62). NFR. D. 1914. Buried in St. John's Cem., Hampton, VA.

WOOD, WILLIAM C.: Co. F & D. 3rd Sgt. Enl. 5/20/61, Williamsburg. 3rd Sgt., 5/20/62. Absent sick 7/12/62. Transferred to Co. D, 10/6/62.

WOODWARD, JOHN R.: Co. C-2. Pvt. Enl. 5/13/61, Richmond; age 32, engineman. AWOL (10/13/61), fined $30. Discharged 8/13/62.

WOOTON, GEORGE D.: Co. C-2. Pvt. Enl. 5/13/61, Richmond; age 42, lawyer. Discharged as armorer, 10/30/61.

WRAY, JACOB K.: Co. A. Pvt. Enl. 3/27/62, Yorktown. Present through (4/62).

WRIGHT, JAMES: Co. G. Pvt. B. York Co. Enl. 5/18/61, Upper Grafton. Absent on leave in York Co., 1/24/62. Discharged 7/15/62; age 45, 5'10", light comp. and hair, grey eyes, sailor.

WRIGHT, JAMES D.: Co. F. Pvt. Enl. 5/20/61, Williamsburg. AWOL 5/6/62. NFR.

WRIGHT, JOHN, Jr.: Co. G. Pvt. Enl. 5/18/61, Upper Grafton. Present through (4/62). NFR.

WRIGHT, WILLIAM: Co. G. Pvt. B. York Co. Age 40, 5'10", farmer, light comp., eyes and hair. Enl. 5/18/61, Upper Grafton. Present through (4/30/62).

WYATT, JOHN WEST: Co. ?. D. Washington, DC on 5/12/1924.

WYVILL, SAMUEL W.: Co. C-1. Pvt. Enl. 7/1/61, Camp Stevens. NFR.

YOUNG, EDWARD D.: Co. C-1. Pvt. Enl. 6/1/61, Manassas. NFR.

YULEE, CARL M.: Co. C-1. Pvt. Enl. 5/1/61, Manassas. Discharged for providing substitute, 10/1/61.

ZIMMERMAN, PURNELL P.: Co. C-1. Pvt. Enl. 5/1/61, Manassas. NFR.

Wise Artillery
James A. Brown's Battery

The Wise Artillery was originally formed in the fall of 1859 in Berkeley County, in what is now West Virginia, in response to John Brown's Raid on Harper's Ferry. A roster of the Wise Artillery, dated October 1859 lists 70 men.

ALBURTUS, EPHRAIM G.: Captain
ALBURTUS, SAMUEL:
BEARD, GEORGE W.:
BLAKELY, EDWARD F.:
BLANCHFIELD, JOHN:
BOWERS, JOHN A.:
BRASIAS, S. H.:
BUCHANAN, JOHN C.: 4th Cpl.
CAGE, RICHARD:
CHAMBERS, GEORGE W.:
CHAMBERS, JAMES H.: 2nd Sgt.
CONWAY, WILLIAM:
COUCHMAN, JACOB:
DORSEY, E. S.:
FEANAN, JAMES D.:
FISK, CHRISTOPHER:
GROVE, DAVID:
GUITMAN, DAVID:
HELFERSTRAY, WILLIAM:
HESS, AARON:
HILL, CHRISTOPHER:
HILL, MARTIN LUTHER:
HINES, J. M.:
HOGAN, FULLWELL:
KEARNES, CYRUS:
KEARNES, ROBERT:
KING, GEORGE H.:
KING, O. M.: 1st Cpl.
LANDERS, MICHAEL:
LANTZ, JOSEPH:
LOWERY, WILLIAM:
LOWMAN, JOHN H.:
LOWRY, ROBERT: 3rd Sgt.
LUCAS, BENJAMIN:
LUCAS, JOHN:
MAHONEY, PATRICK:
MAXWELL, JOHN: 4th Sgt.
McGARY, P. J.:
MIREPER, GEORGE H.: 2nd Lt.
MOORE, JOSEPH:
MURRAY, PATRICK:
O'NEAL, JOHN R.:
PALMER, BENJAMIN:
PENDLETON, PHILIP C.:
PRIOR, THOMAS:
PYRRAL, FREDERICK:
REED, JAMES F.:
REES, JOHN:
ROBINSON, JOHN:
SCHAD, WILLIAM:
SHARY, ADOLPH:

SHAWERS, PHILIP:
SHEA, JOHN:
SISCO, JOHN:
SISCO, PETER:
SMITH, F. T.: 1st Lt.
STEWART, B. W.: 2nd Cpl.
STEWART, CHARLES: Musician (Fifer)
STEWART, JACOB: Musician (Drummer)
SULLIVAN, FREDERICK:
SUTTON, DANIEL: 3rd Cpl.
THOMAS, BARTHOLOMEW:
WALKER, E. M.:
WALLET, GEORGE: 1st Lt.
WALLETT, WILLIAM:
WESTENHAVER, DAVID:
WESTPHAL, CHARLES:
WINCE, HENRY:
WITHROW, W. P.:
ZEPP, GEORGE W.:

Roster, Wise Artillery

ALBURTIS, EPHRAIM GAITHER: Captain. Enl. on ?. Commissioned in militia in 10/1859. Confirmed by election on 11/19/1859 at Martinsburg. Resigned on 1/25/62. On 1/2/63 wrote to the quartermaster general "Having served in the Army as a Capt. of Artillery, until forced to retire by reason of impaired health, and being now compelled to leave my home and family within the lines of the enemy; I beg leave to tender my services as a Clerk in your Dept. if a vacancy exists. I am not subject to the Conscript Law. For reference refer to Maj. Jno. B. Hoge, Messes. Darnell and Dugan of your Dept." B. 7/6/1817 D. 3/21/1875. Buried in the Green Hill Cem., Martinsburg, WV.

ALBURTIS, SAMUEL: Pvt. Enl. on 4/19/61 at Martinsburg, WV. Discharged on surgeon's certificate on 8/16/61 for chronic rheumatism. B. 7/9/1820 D. 8/16/1865. Buried in the Green Hill Cem., Martinsburg, WV.

AMOS, JAMES W.: Pvt. Enl. on 3/31/62 at Orange C.H. Pres. on 3/31/62 and 6/30/62 rolls. Transferred to Taylor's Battery on 10/8/62. Served the remainder of the war. Applied for a pension in May 1900 then living at Somerset, Orange Co., VA. Widow applied for pension on 3/17/1903 in Orange Co.

ARDELLA, J. A.: Pvt. Enl. on ?. "Said to have been discharged" per 3/31/62 roll.

ARMPRIEST, JEROME B.: Pvt. Enl. on 8/26/61 at Martinsburg. Pres. on 2/28/62. In Charlottesville Gen. Hosp. 3/31-4/30/62 with anasarca. Absent, sent to hosp. on 3/31/62. Discharged for disability on 6/12/62. Described as 6', fair comp., grey eyes, light hair, occ. farmer, age 18, born in Berkeley Co.

AULD, CHARLES S.: Pvt. Enl. on 4/19/61 at Martinsburg. Pres. on 8/31/61, 2/28/62, and 3/31/62. Transferred to Maryland Line on 6/5/62, G. O. #38 and S.O. #109, served in Co. A, 1st MD Cav.

BARNES, ANDREW JACKSON: Postwar reference. Williamsburg p.l.

BARNES, L.F.: Pvt. From New Kent Co. Postwar reference.

BAUGHN, GEORGE DALLAS: Postwar reference. Surrendered at Appomattox.

BEALL, RICHARD H.: 5th Sgt. Enl. on 4/19/61 at Martinsburg. Pres. 7-8/61, 1-2/62 as Pvt. and 5- 6/62 rolls as 5th Sgt. Transferred to Taylor's Battery on 10/8/62. Promoted to Sgt. 11- 12/62. Present through the war. Died on 6/14/893, age 47. Buried in the Riner Cem., Hedgesville, Berkeley Co., WV.

BEARD, GEORGE W.: 4th Sgt. Enl. 4/19/61 at Martinsburg. Pres. 7-8/61, absent on 34 day furlough from 2/14/62 as Pvt. Pres. on 3/31/62. Absent 5-6/62, WIA on 6/27/62, sent to General Hosp. 6/27/62. Transferred to Taylor's Battery on 10/8/62. Pres. until WIA again at Bermuda Hundred 12/19/64, sent to Richmond Hosp. 12/23/64. Paroled at Appomattox C.H., VA on 4/9/65. Res. of Berkeley Co., WV. Buried in the Old Norbourne Cem., Berkeley Co., WV.

BECHTEL, JOHN M.: Pvt. Enl. on 9/17/61 at Martinsburg (2/7/62 at Centerville). Pres. on 2/28/62 and 3/31/62. Transferred to Co. A, 1st MD Cav. on 6/5/62 per S.O. #107. Paroled at Winchester, VA on 6/15/65, age 22.

BLACKENEY, DAVID H.: Pvt. Enl. on 4/19/61 at Martinsburg. Pres. 7-8/61, 1-2/62, 2-3/62, and 5- 6/62 rolls. WIA at Sharpsburg, MD on 9/19/62. Transferred to Taylor's Battery on 10/8/62 on paper, never RTD. Conflicting reports indicate extensive hospitalization or desertion.

BLANCHFIELD, JOHN: Pvt. Enl. on 6/13/61 at Martinsburg. Admitted to Chimborazo Hosp. #3 10/22/61 with disease of heart. Discharged on 11/1/61 for carditis. "Said to have been discharged at Richmond by order Secy. of War." per 3/31/62 roll. Shenandoah Co. p.l. D. 1898.

BOYER, JOHN A.: Pvt. Enl. on 4/19/61 at Martinsburg. Pres. on 8/31/61. Admitted to Chimborazo Hosp. #3 10/22-11/1/61 for debility following pneumonia. Absent sent to Gen. Hosp. on 10/20/61, said to have been discharged according to 2/28/62 roll. Actually discharged from service on 11/1/61 for debility. Buried in the Old Norbourne Cem., Berkeley Co., WV.

BRITTON, EDWARD P.: Pvt. Enl. on 6/15/61 at Bunker Hill (2/14/62 at Centreville). Pres. on 7-8/61 due government $4.30 for one blanket and one pair of shoes. Pres. on 1-2/62, 2-3/62, and 5-6/62 rolls. Transferred to Taylor's Battery on 10/8/62. Pres. until he went AWOL 1-2/64.

BROWN, JAMES STEPHENSON: Captain. Elected 1st Lt. on 11/23/1860 at Martinsburg. Pres. commanding Co. on 8/31/61 as 1st Lt. Promoted to Capt. on 2/1/62. Pres. on 2/28/62. Reelected Capt. on 4/26/62. Pres. on 6/30/62. Tendered resignation as Capt. of Artillery, P.A.C.S. on 5/7/64, due to age, then 50. d. 12/14/1886, Orange Co., VA. Age 72. Buried in Graham Cem., Orange Co.

BRYCE, JOHN A.: Pvt. Enl. on ?. "Said to have been discharged at Richmond by order Sec. of War." per 3/31/62 roll.

CAWTHORNE, THOMAS W.: Pvt. Enl. on 2/21/62 in Hanover Co. Transferred to Taylor's Battery on 10/8/62. Pres. or accounted for thru the war.

CHAMBERS, GEORGE H.: Pvt. Enl. on 4/19/61 at Martinsburg. Pres. 7-8/61, 1-2/62, 3/62, and 5-6/62 rolls. Transferred to Taylor's Battery on 10/8/62. Pres. until MWIA at Spotsylvania C.H., VA on 5/12/64, sent to hosp. Died at Camp Winder Gen. Hosp. 6/20/64 from a gunshot wound in the neck and malaria.

CHAPMAN, GEORGE SUEL: Pvt. Enl. on 10/3/62 in Berkeley Co., WV. Transferred from Page's Co. Transferred to Taylor's Battery on 10/8/62. Left sick at Bunker Hill on 7/15/63 on retreat from PA. Taken POW on 7/23/63, sent to Baltimore, MD, to Fort McHenry and then to Point Lookout, MD where held until exchanged on 2/10/65.

CLARK, JOHN WILLIAM: Pvt. B. 2/43. From Orange Co. Enl. on 3/31/62 at Orange C.H., VA. Pres. on 3/31/62. WIA on 6/28/62 at Fair Oaks. Pres. on 6/30/62. Transferred to Taylor's Battery on 10/8/62. Pres. thru 2/28/65. Taken POW on 4/6/65 at Harper's Farm, sent to Point Lookout, MD where held until released on oath on 6/10/65. Res. of Orange Co., 5'7 3/4", florid comp., dark brown hair, light blue eyes. Alive in Unionville, Orange Co., Va., 1915. B. 2/9/1843 at Unionville, Orange Co. D. 10/10/1928 buried in the Rhoadesville Baptist Church Cem., Orange Co., VA.

CLARK, WILLIAM OSCAR: Pvt. Enl. on 3/31/62 at Orange C.H., VA. Pres. on 3/31/62. WIA, small piece of shell in the back, at Chafin's Farm near Richmond, VA 6/62. Pres. on 6/30/62. Transferred to Taylor's Battery on 10/8/62. Absent sick 10/31/62. Discharged for disability by 4/30/63. Applied for pension in 4/1888, age 54. Res. near Unionville, Orange Co. Entered Old Soldier's Home, Richmond on 6/26/1899, age 67. Sent to Western State Asylum, Staunton 6/28/1902, died there on 5/13/1905, age 65. Buried in Unionville, Orange Co., VA.

CLARKE, WILLIAM H.: Pvt. Enl. on 5/18/61 at Harper's Ferry. Pres. on 8/31/61. Absent on furlough on 2/4/62 for 34 days.

CLARKE, WILLIAM J.: Pvt. Enl. on 2/7/62 at Centreville. Pres. on 3/31/63. Transferred to Co. A, 1st MD Cav. by S.O. #107, 6/5/62. In Chimborazo Hosp. #4 8/12-16/62 with injury to the spine. Paroled at New Market, VA on 4/19/65, age 23.

CUMISKEY, THOMAS: Pvt. Enl. on 5/30/61 at Harper's Ferry. Pres. on 8/31/61, 2/28/62, and 3/31/62 rolls. Discharged by conscript act on 4/16/62. D. 3/2/1899, age 75. Buried in the St. Joseph's Catholic Church Cem., Martinsburg, WV.

COSIMINI, GIACINTO: Pvt. Enl. on 3/1/62 in Berkeley Co. (Centreville). Pres. on 3/31/62 and 6/30/62. Transferred to Taylor's Battery on 10/8/62. Died at Rogersville, TN hosp. 8/20/63 with liver cancer exhibiting large abscess.

COUCHMAN, JACOB R.: QM Sgt. Enl. on 4/19/61 at Martinsburg. Pres. on 8/31/61 as Cpl. Appointed 5th Sgt. on 1/1/62. Pres. on 2/28/62 as QM Sgt. Pres. on 3/31/62 as 3rd Sgt. Pres. on 6/30/62 as Pvt. Transferred to Taylor's Battery on 10/8/62. Pres. or accounted for the remainder of the war. Paroled on 4/19/65 at Winchester, age 28, 5'11", fair comp., fair hair, hazel eyes. B. 1837 D. 1885, buried in the Green Hill Cem., Martinsburg, WV.

COWHERD, FRANK: Pvt. From Orange Co. Enl. on 4/5/62 at Orange C.H. Pres. on 3/31/62. Absent on detached service as QM on 6/30/62. Paid for service from 3/31/62 to 10/8/62.

COX, SAMUEL S.: Pvt. Enl. on 8/26/61 at Martinsburg. Pres. on 1-2/62 roll. Pres. on 3/31/62 and 6/30/62 rolls. Transferred to Taylor's Battery on 10/8/62. Deserted at Bunker Hill, Berkeley Co., WV on 7/14/63.

CUNNINGHAM, DAVID: Pvt. Enl. on 4/19/61 at Martinsburg. Pres. on 8/31/61, 2/28/62, and 3/31/62 rolls. Discharged by conscript act on 4/16/62.

DAVIS, JOHN: Pvt. From Orange Co. Enl. on 3/31/62 at Orange C.H. Pres. on 3/31/62 and 6/30/62. Transferred to Taylor's Battery on 10/8/62. Died of typhoid fever on 8/1/63 at Gordonsville, VA Hosp.

DAVIS, RICHARD R.: Pvt. Enl. on 3/31/62 at Orange C.H. Pres. on 3/31/62 and 6/30/62 rolls. Transferred to Taylor's Battery on 10/8/62. Sick on 10/31/62 roll. Pres. thru 2/28/65. Paroled at Appomattox C.H., VA on 4/9/65. Res. of Burr Hill, Orange Co., VA in 7/1902, then age 74.

ELLIOTT, WILLIAM T.: Pvt. Enl. on 3/31/62 at Orange C.H. Pres. on 3/31/62. Absent sick in Gen. Hosp. per 6/30/62 roll. Sick in Chimborazo Hosp. #3 with typhoid fever 7/2/62. Transferred to Taylor's Battery on 10/8/62. Absent sick on 10/31/62. Pres. thru 2/28/65. Taken POW at Harper's Farm, Amelia Co. on 4/6/65, sent to Point Lookout, MD where held until paroled on 6/11/65. Res. of Greene Co., 5'11-", light comp., light hair, dark blue eyes.

ESTES, ISAAC M. N.: Pvt. Enl. on 3/31/62 at Orange C.H. Pres. on 3/31/62 and 6/30/62 rolls. Transferred to Taylor's Battery on 10/8/62. Absent sick 10/31/62. Died on 3/5/63 at Mr. Edward P. Woods' in Caroline Co.

FAULCONER, BENJAMIN FRANKLIN: Pvt. Enl. on 3/31/62 at Orange C.H. Pres. on 3/31/62 and 6/30/62 rolls. In Chimborazo Hosp. #4 8/18/62 to 9/9/62 for continued fever. Furloughed 9/9-10/14/62. Discharged on 8/11/62 for general condition and tubercular consumption, age 42, 5'10", fair comp., brown hair, blue eyes. B. 1822 in Orange Co. D. 1874, buried in the Massey Family Cem., Orange Co., VA.

FAULCONER, JOHN THORNTON: Pvt. Enl. on 3/31/62 at Orange C.H. Pres. on 3/31/62 and 6/30/62 rolls. Transferred to Taylor's Battery on 10/8/62. Pres. 10/31/62. Detailed wagoner and in QM Dept. much of the war. Paroled at Appomattox C.H., VA on 4/9/65. B. 3/5/1832 in Culpeper Co., VA. D. 1/17/1913 in Culpeper Co.

FAULKNER, ELISHA BOYD: Pvt. B. 7/24/41. Enl. on 6/10/61 at Point of Rocks. Pres. on 8/31/61. Also served with 1st Rockbridge Artillery. Alive in 1913.

FLETCHER, MADISON: Pvt. Enl. on 3/31/62 at Orange C.H., VA. Pres. on 3/31/62. Hospitalized on 5/11/62. In Danville Gen. Hosp. 6/4/62 with debilities. Discharged on 10/4/62, carried on rolls of Taylor's Battery, however, until he was listed as having died at home by 6/30/64. Res. of Gordonsville, Orange Co., VA.

FLETCHER, RAY: Postwar reference. From Orange Co.

FRAZIER, JAMES W. B.: Pvt. Enl. on 4/19/61 at Martinsburg, VA. Pres. 8/31/61, 2/28/62, 3/31/62 and 6/30/62 rolls. Transferred to Taylor's Battery on 10/8/62. Taken POW at Shepherdstown (or Martinsburg) on retreat from PA by Ring Cav., sent to Wheeling and then to Camp Chase. Described as res. of Haysville, Berkeley Co., occ. railroad hand, age 22, 6', dark comp., black hair, blue eyes. Transferred to Camp Chase, OH on 7/22/63 and then to Fort Delaware on 3/4/64. Released on oath on 6/11/65.

FULTZ, JOHN HAMPDEN: Pvt. Enl. on 5/20/62 at Harper's Ferry, VA (2/7/62 at Centerville). Pres. on 8/31/61. Absent on 34 day furlough on 2/14/62. Pres. on 3/31/62. Transferred to Co. E, 1st MD Cav. by S.O. 107, dated 6/5/62.

FUNNUCAN, JOHN P.: Pvt. Enl. on 6/13/61 at Harper's Ferry. Pres. on 8/31/61. Deserted 1/9/62. RTD on 2/11/62. Pres. on 2/28/62. Pres. 3/31/62. Discharged by conscript act on 4/16/62.

GALLAGHER, FRANCIS: Pvt. Enl. on 6/21/61 at Winchester. Pres. on 8/31/61. Absent 2/28/62. Sent to Warrenton Moore Hosp. 1/22/62. and 4/23/62 with mucous diarrhea. Pres. on 6/30/62. Transferred to Taylor's Battery on 10/8/62. Deserted in Maryland ca. 9/12/62. Released on oath, to remain in any of the "loyal states." B. in Ireland. Age 45 in 1870. Buried in the St. Joseph's Catholic Church Cem., Martinsburg, WV.

GALLAGHER, PATRICK: Pvt. Enl. on 7/6/61 at Darkeville, VA. Pres. on 8/31/61, 2/28/62, and 3/31/62. Discharged on 4/16/62 by conscript act. D. 5/30/1885, buried in the St. Joseph's Catholic Church Cem., Martinsburg, WV.

GEABKE, (GAEBKE) THEODORE: Pvt. Enl. on 8/6/61 at Winchester, VA. Pres. on 8/31/61, 2/28/62, 3/31/62. Discharged by conscription act on 4/16/62.

GENTRY, SILAS C.: Pvt. Enl. on 3/31/62 at Orange C.H., VA. Pres. on 3/31/62 and 6/30/62 rolls. In Chimborazo Hosp. #2 8/13-9/30/62 with continued fever. Transferred to Taylor's Battery on 10/8/62. Reported of deserters arrested by 1st Lt. F. M. McMullan, Co. D, 4th VA Heavy Art. at Camp Wappoo, SC on 10/31/63. Other records indicate that he was left sick at Cashtown, MD on 7/5/63. Still other records indicate that he was left WIA in the wrist at Gettysburg, PA, and taken POW, sent to Fort Delaware where he joined Co. G, 1st U.S. Vol. Inf. or Co. F, 2nd Connecticut Heavy Art. Received U.S. Invalid pension and his widow received a pension in 1917.

GRAVES, JACOB C.: Pvt. Enl. on 4/5/62 at Orange C.H. Left sick in Orange Co. on 4/6/62, never rejoined.

GRAVES, RICHARD PERRIN: 1st Lt. From Orange Co. Enl. on 3/31/62 at Orange C.H. Elected 1st Lt. on 4/26/62. Resigned on 11/10/62.

GRAVES, WILLIAM C.: 3rd Lt. From Orange Co. Enl. on 3/31/62 at Centreville. Elected 3rd Lt. on 4/26/62.

HARDY, GEORGE W.: Pvt. Enl. on 8/26/62 at Martinsburg (2/7/62 at Centreville, VA.) Hospitalized on 2/16/62. Died on 3/8/62 of typhoid fever.

HARLEY, JAMES: Pvt. Enl. on 6/12/61 at Harper's Ferry. Pres. on 8/31/61, 2/28/62, 3/31/62 and 6/30/62 rolls. Transferred to Taylor's Battery on 10/8/62. Pres. until KIA by Federal guns in the road near Chancellorsville on 5/12-13/63.

HAZARD, CHARLES: Pvt. Enl. on 8/14/61 at Manassas (2/12/62 at Centerville). Pres. on 8/31/61, 2/28/62, 3/31/62 and 6/30/62 rolls. Transferred to Taylor's Battery on 10/8/62. MWIA in the bladder at Cold Harbor on 6/1/64. Died on 7/7/64 at Louisiana Hosp. #1, Richmond, VA.

HEDGES, BAILEY S.: Pvt. Enl. on 6/15/61 at Bunker Hill. Pres. on 8/31/61, 2/28/62, 3/31/62, and 6/30/62 rolls. Transferred to Taylor's Battery on 10/8/62. Pres. thru 2/28/65. Paroled at Appomattox C.H., VA on 4/9/65. B. 5/29/1839. D. 5/26/1906, buried in the Green Hill Cem., Martinsburg, WV.

HELFERSTAY, WILLIAM: Pvt. Enl. on 4/19/61 at Martinsburg, VA. Pres. on 8/31/61, 2/28/62, and 3/31/62 rolls. Discharged by conscript act on 4/16/62. B. 6/29/1829 D. 6/24/1898. Buried in the Green Hill Cem., Martinsburg, WV.

HERNDON, JOHN Q.: Pvt. Enl. on 3/31/62 at Orange C.H. Pres. on 3/31/62. In Hosp. per 6/30/62 roll, WIA at Mechanicsville on 5/24/62. Died in service, J. L. Woolfolk, attorney, filed claim for pay and allowances on 12/10/64. Born in Orange Co.

HERNDON, THOMAS H.: Pvt. Enl. on 5/23/61 at Harper's Ferry, VA. (2/21/62 at Centreville). Pres. on 8/31/61. Hospitalized in Chimborazo Hosp. #3 11/22/61 to 1/20/62 with remittent fever. Pres. on 2/28/62 and 3/31/62. Transferred to Co. A, 1st MD Cav. on 6/5/62 by S.O. 107. Paroled at New Market, VA on 4/20/65, age 28.

HERNDON, W. W.: Pvt. Enl. on 3/31/62 at Orange C.H., VA. Pres. on 3/31/62. Sent to Gen. Hosp. #21, Richmond on 5/10/62 with rubeola. Died on 5/25/62.

HESS, AARON T.: Pvt. B. 9/6/40. Enl. on 4/19/61 at Martinsburg. Pres. on 8/31/61 roll. Absent on 2/28/62 roll. Pres. on 3/31/62 and 6/30/62 rolls. Transferred to Taylor's Battery on 10/8/62. Taken POW at Harper's Farm, Amelia Co., VA on 4/6/65, sent to Point Lookout, MD where held until released on 6/13/65. Res. of Berkeley Co., WV, 5'6", fair comp., brown hair, light blue eyes. (Diary in possession of family). D. 9/3/1922, buried in the Green Hill Cem., Martinsburg, WV.

ISRAEL, GILBERT T.: Pvt. Enl. on 8/25/61 at Winchester (2/7/62 at Centreville). Pres. on 8/31/61. Absent on furlough on 2/14/62 for 34 days. Pres. on 3/31/62 and 6/30/62 rolls. Transferred to Taylor's Battery on 10/8/62. Pres. thru 2/28/64, transferred to the 2nd MD Artillery on 4/21/64 by Special Orders 93/2.

JACOBS, BENJAMIN: Pvt. Enl. on 8/25/61 in Winchester (4/7/62 at Orange C.H.). Pres. on 8/31/61. Absent on furlough on 2/14/62. Pres. on 3/31/62. Hosp. 5/11/62, still absent on 6/30/62. Transferred to Taylor's Battery on 10/8/62. AWOL 9-12/63 and 1-2/64. Taken POW on 1/10/64 on march from Loudoun to Knoxville, TN. Sent to Camp Chase, OH on 2/4/64, then to Rock Island, IL where held until he died. Buried in Grave 525, Confederate Cem., Rock Island, IL.

JARDELLA, J. ANDRIA: Pvt. Enl. on 5/29/61 at Harper's Ferry. Pres. on 8/31/61. Absent in Gen. Hosp. on 10/20/61. Discharged on 11/1/61 for carditis.

JOHNSON, ISAAC H.: Pvt. Enl. on 5/18/61 at Harper's Ferry, VA. Pres. on 8/1/61. Sent to Hosp. on 11/20/61. Discharged for disability on 6/12/62, described as 6'2", age 22, fair comp., hazel eyes, light hair, occ. farmer. Born in Harford Co., MD.

KERNS, CYRUS McLEAN: 2nd Lt. Enl. on 4/19/61 at Martinsburg (2/7/62) at Centreville, VA). Pres. on 8/31/61 as Cpl. Absent recruiting by order of Gen. Johnston on 2/26/62 as 4th Sgt. Reelected 2nd Lt. on 4/26/62. KIA on 6/28/62 at Garnett Farm.

KEYS, JOHN T.: Pvt. Enl. on 5/15/61 (5/30/61) at Harper's Ferry. Transferred from Co. D, 1st VA Art. Pres. on 8/31/61. In Chimborazo Hosp. #3 for dyspepsia. RTD 12/18/61. Sent to Hosp. 2/27/62, attached as a nurse 5/9/62. Transferred to Taylor's Battery on 10/8/62. Pres. at hosp. most of the war, recommended for clerkship as incapable of performing active duty, plural adhesions from attacks of pneumonia, pleuritis, and partial paralysis of the left arm. Applied on 3/31/63 to Sec. of the Treasury as clerk. Taken POW at Chester Gap, Fauquier Co. on 7/21/63, furloughed on 7/2662 by Buford's Cav. Transferred to Point Lookout, MD, exchanged on 4/27/64. Pres. 6/30/64. Admitted to Chimborazo #3 for debility on 5/1/64. Paroled at Appomattox C.H., VA on 4/9/65. Res. of Fauquier Co., VA.

KINZER, GEORGE W.: Pvt. Enl. on 8/21/61 at Winchester. Pres. on 8/31/61, 2/28/62, and 3/31/62. Deserted before the enemy on 7/2/62.

KINZER, PHILIP F.: Pvt. Enl. on 3/31/62 at Orange C.H., VA. Previously served as 1st Lt., Orange Co. Home Guard formed on 4/22/61. Pres. on 3/31/62. Detailed as a teamster on 4/19/62. Pres. on 6/30/62. Absent sick on 10/31/62. Transferred to Taylor's Battery on 10/8/62. Pres. thru 2/65 roll. Paroled at Appomattox C.H., VA on 4/9/65. Living in Orange Co., VA in 1870, age 51. Died on 11/28/1882. Widow drew pension under act of 1900.

KINZER, WILLIAM H.: Pvt. Enl. on 8/21/61 in Winchester. Pres. on 8/31/61, 2/28/62, and 3/31/62 rolls. Deserted before the enemy on 7/2/62.

KNIGHTON, ROBERT T.: Pvt. Enl. on 4/5/62 at Orange C.H., VA. Pres. on 3/31/62 and 6/30/62 rolls. Transferred to Taylor's Battery on 10/8/62. Absent sick 10/31/62. Absent sick most of the war.

LANCASTER, OWEN C.: Pvt. Enl. on 3/31/62 at Orange C.H., VA. Pres. on 3/31/62 and 6/30/62 rolls. Transferred to Taylor's Battery on 10/8/62. AWOL on 10/31/62 roll. RTD by 2/28/63. Taken POW on Mossy Creek, TN on 1/22/64 while on forage detail. Sent to Camp Chase, OH and then to Rock Island, IL where held until exchanged on 3/2/65. Living at St. Just, Orange Co. in 1870 age 47. Applied for pension in 1900 then age 77. Alive in 5/1902.

LANCASTER, R. S.: Pvt. Enl. on 3/31/62 at Orange C.H. Pres. on 3/31/62. WIA at Mechanicsville, VA on 5/20/62. Transferred to Eubank's Battery. Retired to Invalid Corps. on 12/30/64.

LANCASTER, WILLIAM C.: Pvt. Enl. on 3/31/62 at Orange C.H., VA. Pres. on 3/31/62 and 6/30/62 rolls. Transferred to Taylor's Battery on 10/8/62. Sick on 10/31/62 roll RTD by 10/31/63. WIA before 6/62. In Chimborazo Hosp. #3 6/21/64 with chronic diarrhea. Died of disease at Lynchburg Gen. Hosp. 7/28/64. Res. of Mine Run, Orange Co., VA. Widow applied for pension in 1888.

LANDERS, MICHAEL: Pvt. Enl. on 4/19/61 at Martinsburg. Pres. on 8/31/61. Reported discharged on 12/25/61. Pres. on 2/28/62 and 3/31/62. Discharged by conscript act on 4/16/62.

LANGLEY, J. LYLE: Bugler. Enl. on 8/26/61 at Winchester, VA. Absent sick leave for 30 days beginning on 1/28/62. Pres. on 3/31/62. Discharged by conscript act on 4/16/62. Entered R.E. Lee Camp Soldier's Home Richmond, 5/28/93, age 72. Died 6/22/02. Buried Hollywood Cem., Richmond, Va.

LANTZ, CHRISTIAN: Pvt. Enl. on 3/8/62 in Berkeley Co. (Winchester, VA). Pres. on 3/31/62 and 6/30/62 rolls. Transferred to Taylor's Battery on 10/8/62. Pres. thru 2/28/65. Paroled at Appomattox C.H., VA on 4/9/65. Res. of Berkeley Co. Buried in the Green Hill Cem., Martinsburg, WV.

LANTZ, JOSEPH T. VAN L.: 1st Cpl. Enl. on 4/19/61 at Martinsburg. Pres. 8/31/61. Absent on 34 day furlough on 2/14/62. Pres. on 3/31/62 and 6/30/62 rolls. Transferred to Taylor's Battery on 10/8/62. KIA at Gettysburg, PA. Buried in the Green Hill Cem., Martinsburg, WV.

LEE, SYLVESTER SAMUEL: Pvt. Enl. on 3/31/62 at Orange C.H. Pres. on 3/31/62 and 6/30/62 rolls. Transferred to Taylor's Battery on 10/8/62. Pres. thru 2/28/65. Paroled at Appomattox C.H., VA on 4/9/65. Res. of Orange Co. Died there on 8/13/1901. Occ. railroad worker.

LEIGHTER, CHARLES W.: Pvt. Enl. on 9/17/61 at Martinsburg. (2/7/62 at Centreville.) Pres. on 8/31/61. Pres. on 3/31/62. Transferred to Co. A, 1st MD Cav. on 6/5/62 by S.O. 107.

LEWIS, PHILIP S.: Pvt. Enl. on 3/31/62 at Orange C.H. Pres. on 3/31/62. Sent to Hosp. on 6/20/62. Left home sick when the company marched for Richmond. Transferred to Taylor's Battery on 10/8/62. AWOL on 12/31/63 roll. Taken POW on 12/3/63 at Russellville, TN, sent to Camp Chase, OH and then to Rock Island, IL where he died of disease on 3/19/64 of chronic diarrhea. Buried in grave #857 in the Confederate Cem., Rock Island, IL. Letters to wife in USAMHI and in the Virginia Historical Society.

LUCAS, BENJAMIN: 2nd Sgt. Enl. on 4/19/61 at Martinsburg. Pres. on 8/31/61 roll. On 34 day furlough 2/14/62. Pres. on 3/31/62 as 2nd Sgt. Pres. on 6/30/62 as Pvt. Transferred to Taylor's Battery on 10/8/62. Listed as AWOL on 10/31/62 and subsequent muster rolls. B. 11/5/1817 D. 4/7/1890. Buried in the Edge Hill Cem., Charles Town, WV.

LUCAS, CHARLES W.: 4th Cpl. Enl. on 4/19/61 at Martinsburg (3/11/62 at Centerville.) Pres. on 8/31/61 and 2/28/62 rolls as pvt. Pres. on 3/31/62 as Cpl. KIA at Garnett Farm on 6/27/62.

LUCAS, OSCAR M.: Pvt. Enl. on 4/19/61 at Martinsburg (3/8/62 at Centreville). Pres. on 8/31/61. On 34 day furlough on 2/14/62. Pres. on 3/31/62 and 6/30/62 rolls. Transferred to Taylor's Battery on 10/8/62. Pres. on all subsequent muster rolls. Paroled at Appomattox C.H., VA on 4/9/65. Res. of Jefferson Co., WV. D. 5/22/1888, buried in the Edge Hill Cem., Charles Town, WV.

MAHONEY, PATRICK: Pvt. Enl. on 4/17/61 at Martinsburg. Pres. on 8/31/61, 2/28/62, and 3/31/62 rolls. Discharged by conscript act on 4/16/62. B. 1818 D. 4/2/1886. Buried in the St. Joseph's Catholic Church Cem., Martinsburg, WV.

MARKLE, SAMUEL T.: Pvt. Enl. on 6/10/61 at Harper's Ferry. Pres. on 8/31/61, 2/28/62, 3/31/63 and 6/30/62 rolls. Transferred to Taylor's Battery on 10/8/62. Pres. thru 2/28/65. Promoted to Cpl. 9-12/63. Paroled at Appomattox C.H., VA on 4/9/65. Res. of Berkeley Co., WV. B. 6/4/1810 D. 2/3/1913. Buried in the Episcopal and Masonic Cem., Jefferson Co., WV.

MASTON, ELIJAH J.: Pvt. Enl. on 3/31/62 at Orange C.H. Pres. on 3/31/62 and 6/30/62 rolls. Detailed as a teamster in 4/62. Transferred to Taylor's Battery on 10/8/62. Detailed as a wagoner or teamster until taken POW on 1/20/64 at Mossy Creek, TN. Sent to Camp Chase, OH and then to Rock Island, IL, where held until he died on 3/29/64, buried in grave 947, Confederate Cem.

MAXWELL, JOHN: Brevet 2nd Lt. Enl. on 4/19/61 at Martinsburg. Pres. on 8/31/61 as Sgt. Pres. on 2/28/62 as 2nd Lt. Failed to be reelected, dropped on 4/26/62. Entered R.E > Lee Soldiers Home, Richmond, Va., 3/10/08, age 75. Dropped at his own request 8/17/16.

McLAUGHLIN, FRANKLIN (FRANCIS) "FRANK": Pvt. Enl. on 6/12/61 at Harper's Ferry. Pres. on 8/31/61 and 2/28/62 rolls. Absent sick on 3/31/62 roll. Discharged by conscript act on 4/16/62. Buried in the St. Joseph's Catholic Church Cem., Martinsburg, WV.

MEANS, T. A.: Asst. Surgeon

MOONEY, JOHN B.: Pvt. Enl. on 5/18/61 at Harper's Ferry. (2/7/62 at Centreville). Absent, sent to Hosp. at Orange C.H. on 8/31/61. Absent on furlough since 2/14/64 for 34 days per 2/28/65 roll. Pres. on 3/31/62. Transferred to Co. A, 1st MD Cav. on 6/5/62 by S.O. 107. Paroled at Winchester, VA on 5/29/65, age 21.

MOORE, ANDREW M.: Pvt. Enl. on 4/19/61 at Martinsburg. Pres. on 8/31/61 and 2/28/62 rolls. Absent sick in hosp. for phthisis pulmonalis on 3/31/62 roll. Died 4/27/1862, buried in the Confederate Cem., University of Virginia, Charlottesville, VA.

MULLIGAN, JOHN PATRICK: Pvt. Enl. on 3/8/62 at Winchester. Pres. on 3/31/62 and 6/30/62 rolls. Samuel Brandt, Sheriff, forwarded him to Capt. Thomas S. McGowan, U.S. Army Provost Marshall, Chambersburg, PA on 9/22/64. B. 9/3/1847 D. 4/17/1893, buried in the St. Joseph's Catholic Church Cem., Martinsburg, WV.

MURPHY, GEORGE H.: 2nd Lt. Enl. on 4/19/61 at Martinsburg. Resigned on 7/30/61 at Manassas, VA.

MURRY, PATRICK: Pvt. Enl. on 4/19/61 at Martinsburg. Pres. on 8/31/61. In Chimborazo Hosp. #4 10/26-11/18/61 with bronchitis. Pres. under arrest on 2/28/62, charged with AWOL. Pres. on 3/31/62 roll. Discharged by conscript act on 4/16/62.

NEWMAN, JOSEPH L.: Pvt. Enl. on 3/20/62 at Orange C.H. Pres. on 3/31/62 and 6/30/62 rolls.

NOLAND, WILLIAM: Pvt. Enl. on 7/27/61 at Manassas (3/4/62 at Centerville). Pres. on 8/31/61. Pres. on 2/28/62, under arrest for desertion on 2/28/62. Pres. on 6/30/62. Transferred to Taylor's Battery on 10/8/62. Detailed as a teamster. Pres. thru 2/28/65. Entered R.E. Lee Camp Soldiers Home, Richmond, Va., 12/5/07, age 80. Dropped on account of AWOL 10/8/08.

NORMAN, G. S.: Surgeon

PALMER, WILLIAM K.: Pvt. Enl. on 4/19/61 at Martinsburg. Pres. on 8/31/61, 2/28/62, 3/31/62 and 6/30/62 rolls. Transferred to Taylor's Battery on 10/8/62. Pres. on all subsequent rolls. Paroled at Appomattox C.H., VA on 4/9/65. Res. of Berkeley Co. D. in 1901, buried in the Green Hill Cem., Martinsburg, WV.

PAYNE, JOHN ROBERT: Postwar reference. B. 1/1/49, New Post, Va. D. 8/1/09, Eggleston, Va.

PELHAM, JOHN: 1st Lt. B. 9/7/38, Calhoun Co., Alabama. Attended USMA. Lt. of Artillery, P.A.C.S., 3/16/61. Served as lieutenant/drill master w/ Alburtis' Battery, 1861. Commanded as 1st Lt., the Newtown Artillery 10/61 - the reorganization of the battery as the nucleus of the Stuart Horse Artillery. Captain of Pelham's Stuart Horse Artillery through 3/23/62. Major 8/9/62. Lt. Col. 4/4/63 (posthumously) and backdated to 3/2/63. MWIA at Kelly's Ford, Va. 3/17/63. Buried in Jacksonville, Alabama.

PIERCE, JOHN W.: Pvt. Enl. on 3/31/62 at Orange C.H. AWOL on 3/31/62 roll. Left sick in Orange Co. on 4/8/62 per 6/30/62 roll.

PRIEST, ALBERT TELLOUS: Pvt. B. 2/23/47. Enl. on 3/31/62 at Orange C.H., VA. Pres. on 3/31/62. Hosp. on 6/27/62 with typhoid. Transferred to Taylor's Battery on 10/8/62. Absent sick on 10/31/62 roll. WIA in the Wilderness 5/5-6/64. Taken POW at Harper's Farm, Amelia Co., VA 4/6/65, sent to Point Lookout, MD where held until released on oath on 6/16/65, res. of Orange Co., 5'8-", light comp., brown hair, hazel eyes. D. 10/7/26. Buried in the Waugh/Priest Fam. Cem., Clark Mtn., Orange Co., VA.

PRYOR, THOMAS E.: Pvt. Enl. on 4/19/61 at Martinsburg, WV. Pres. on 8/31/61, 2/28/62, 3/31/62 and 6/30/62 rolls. Detailed by S.O. 101/3 to work in the Richmond's Tredegar Iron Works.

REARDON, JOHN: Pvt. Enl. on 5/20/61 at Harper's Ferry. Pres. on 8/31/61 and 2/28/62 rolls. Pres. on 3/31/62. Discharged by conscript act on 4/16/62.

RHOADES, JOHN: Pvt. Enl. on 3/31/62 at Orange C.H., VA. Pres. on 3/31/62 and 6/30/62 rolls. Transferred to Taylor's Battery on 10/8/62. Pres. on all subsequent muster rolls. Res. of Greene Co.

RICHARDS, JAMES W.: Pvt. Enl. on 3/31/62 at Orange C.H. Pres. on 3/31/62 and 6/30/62 rolls.

RIORDAN, JOHN D.: Postwar reference. Living at 820 St. Peter St., Baltimore in 1894.

ROACH, JOHN: Pvt. Enl. on 8/26/61 at Martinsburg (2/7/62 at Centreville). Pres. on 2/28/62, 3/31/62 and 6/30/62 rolls. Transferred to Taylor's Battery on 10/8/62. Detailed as a teamster until taken POW at Mossy Creek, TN. Sent to Camp Chase, OH and then to Rock Island, IL where held until he joined U.S. Army for frontier service on 10/17/64.

ROBINSON, JOHN S.: Sgt. Enl. on 4/19/61 at Martinsburg. Pres. on 8/31/61. Discharged on 11/30/61 for a gross defect in his vision. Described as age 46, 5'8", born in England, occ. laborer, fair comp., blue eyes, brown hair. B. 6/26/1819 D. 4/13/1877. Buried in the Green Hill Cem., Martinsburg, WV.

ROSE, AUGUSTUS P.: Pvt. Enl. on 4/27/61 at Harper's Ferry. Transferred to Danville Greys, 18th VA Inf. on 8/4/61 by S.O.

ROUTT, WILLIAM H.: Pvt. Enl. on 3/31/62 in Greene Co. (Orange C.H.) Pres. on 3/31/62 and 6/30/62 rolls. Transferred to Taylor's Battery on 10/8/62. Pres. thru 2/28/65. Age 38, bricklayer, 5', paroled at Stanardsville, VA by 11th PA Cav. (22 letters to his wife at the Museum of the Confederacy.)

SAYER, BENJAMIN FRANKLIN: Postwar reference. Entered R.E. Lee Soldiers Home, Richmond, 9/13/05, age 89. Died 10/7/07. Buried Hollywood Cem., Richmond, Va.

SAWYERS, JOSEPH A.: Postwar reference. B. 3/8/40.

SCHULTZ, WILLIAM: Pvt. Enl. on 4/19/61 at Martinsburg. Pres. on 8/31/61, 2/28/62, 3/31/62 and 6/30/62 rolls.

SEIBERT, JOSEPH: Cpl. Enl. on 5/25/61 at Harper's Ferry. Pres. on 8/31/61 and 2/28/62 as Pvt. Pres. on 3/31/62 and 6/30/62 rolls. Transferred to Taylor's Battery on 10/8/62. Pres. all rolls until listed as AWOL 2/29/64. Buried in the Green Hill Cem., Martinsburg, Berkeley Co., WV.

SHEA, JOHN: Pvt. Enl. on 4/19/61 at Martinsburg. Pres. on 8/31/61. Listed as discharged on 12/12/61. Pres. on 2/28/62 and 3/31/62 rolls. Discharged by conscription act on 4/16/62. D. 7/14/1914, in "his 85th year." Buried in the St. Joseph's Catholic Church Cem., Martinsburg, WV.

SHEIG, ADOLPHUS: Pvt. Enl. on 4/19/61 at Martinsburg. Pres. on 8/31/61, 2/28/62, and 3/31/62. Discharged by conscription act on 4/16/62. B. 9/24/1824 D. 4/28/1883. Buried in the Green Hill Cem., Martinsburg, WV.

SHERRER, (SHERROR) JOSEPH E.: Bugler. Enl. on 6/17/61 at Bunker Hill (2/19/62 at Centerville). Pres. on 8/31/61. Pres. on 2/28/62 as acting bugler. Pres. on 3/31/62 and 6/30/62 rolls. Acting bugler 2/1/62. Transferred to Taylor's Battery on 10/8/62. Pres. all rolls thru 2/28/65. Taken POW at Harper's Farm, Amelia Co., VA on 4/6/65. Sent to Point Lookout, MD where held until released on oath on 6/19/65, 5'5", dark comp., brown hair, blue eyes. Transportation furnished to Martinsburg, WV from Washington, DC. B. 3/10/1830. D. 12/17/1900, buried in the St. Joseph's Catholic Church Cem., Martinsburg, WV.

STEWART, BERNARD "Barney" W.: 1st Sgt. Enl. on 4/19/61 at Martinsburg. Pres. on 8/31/61. Discharged on 2/10/62 for disability (chronic inflamation of the bowels), described as age 38, 5'6", brown eyes, dark comp., dark hair, occ. mechanic. Born in Martinsburg, Berkeley Co., WV. Buried in the Green Hill Cem., Martinsburg, WV.

STRANEY, EDWARD F.: Pvt. Enl. on 6/2/61 at Harper's Ferry. Pres. on 8/31/61, 2/28/62, and 3/31/62 rolls. Hospitalized in Gen. Hosp. #21 on 5/10-17/62 with "old wound." Date and place WIA not noted. Detailed to work at Tredegar Iron Works, Richmond, per 6/30/62 roll per S.O.# 101, 5/2/62. Transferred to Taylor's Battery on 10/8/62. Detailed to Richmond and Petersburg Railroad Company on 5/2/62 by Special Orders 101/3. Absent on all subsequent muster rolls, absent working at Tredegar Iron Works, Richmond, VA.

STRAYER, ADAM P.: Pvt. Enl. on 4/19/61 in Martinsburg, WV. Pres. on 8/31/61 and 2/28/62 rolls. Absent in Chimborazo Hosp. #2 on 3/3-7/62, with syphilis. Discharged by conscript act on 4/16/62.

SUTER, THOMAS C.: Pvt. Enl. on 5/20/61 at Harper's Ferry (2/7/62 at Centreville.) Pres. on 8/31/61. Absent on furlough on 2/14/62 for 34 days per 2/28/62 roll. Absent on furlough "supposed to be in the hands of the enemy" per 3/31/62 and 6/30/62 rolls. No Federal record of incarceration.

SUTHERLAND, ROBERT: Pvt. Enl. on 3/31/62 at Orange C.H., VA. Pres. on 3/31/62 and 6/30/62 rolls. Transferred to Taylor's Battery on 10/8/62. Absent sick on 10/31/62 roll. RTD by 2/28/63. Hospitalized at Marietta, GA per 12/31/63 roll. RTD by 2/29/64. Pres. thru 2/28/65. Taken POW on 4/6/65 in Amelia Co., sent to Point Lookout, MD where held until he died of disease on 6/29/65. Buried in the Confederate Cem., Point Lookout, MD.

SUTHERLAND, WILLIAM: Pvt. Enl. on 3/31/62 at Orange C.H., VA. Pres. on 3/31/62 and 6/30/62 rolls. Transferred to Taylor's Battery on 10/8/62. AWOL on 10/31/62 roll. Pres. on 2/28/63 roll. WIA in leg at Chancellorsville on 5/3/63. Furloughed 15 days per 6/30/64 roll. Pres. thru 2/28/65. Paroled at Appomattox C.H., VA on 4/9/65. Living at True Blue, Orange Co., VA in 5/1900, then age 75, when he applied for a pension.

TANNER, ABRAHAM: Pvt. Enl. on 3/31/62 at Orange C.H., VA. Pres. on 3/31/62 and 6/30/62 rolls. Transferred to Taylor's Battery on 10/8/62. Pres. all rolls thru 12/31/64. Absent on 2/28/65, on 15 day furlough. Taken POW on 4/6/65 in Amelia Co., sent to Point Lookout, MD where held until released on oath on 6/21/65. Res. of Madison Co., 6'-", light comp., brown hair, gray eyes.

TATE, ROBERT J.: 3rd Sgt. Enl. on 6/30/61 at Winchester. Pres. on 8/31/61 as Pvt., on 2/28/62 as 3rd Cpl., on 3/31/62 as 2nd Cpl. Pres. on 6/30/62 as 3rd Sgt. Transferred to Taylor's Battery on 10/8/62. Pres. all subsequent rolls. Paroled at Appomattox C.H., VA on 4/9/65. Res. of Berkeley Co., WV.

TATUM, JOSEPH H.: Pvt. Enl. on 3/31/62 at Orange C.H., VA. Pres. on 3/31/62. Absent in Hosp. sine 6/24/62 per 6/30/62 roll. Transferred to Taylor's Battery on 10/8/62. Sick on 10/31/62 roll. Discharged by 4/30/63 for disability at Mt. Carmel Church, Caroline Co., VA. Res. of Tatum, Orange Co., VA. Applied for pension age 69 in 5/1900. Died 8/1911, Tatum, Orange Co., VA.

TAYLOR, H. C.: Postwar reference. Resident of New Kent Co.

TINDER, GEORGE THOMAS: Pvt. Enl. on 3/31/62 at Orange C.H., VA. Hospitalized 4/28/62. WIA by a shell in the abdomen near Yorktown on 6/24/62. Transferred to Taylor's Battery on 10/8/62. Absent sick on 10/31/62 roll. Detailed to drive ambulance thru 12/31/63. Detailed to drive ordnance train 1/64-2/28/65. Paroled at Appomattox C.H., VA on 4/9/65. Res. of St. Just, Orange Co., VA in 1900 when he applied for a pension, then age 68. D. 1/1/1904 of an inflamation of the bladder.

TINDER, JOHN A.: Pvt. Enl. on 3/31/62 at Orange C.H. Pres. on 3/31/62 and 6/30/62 rolls.

TITLOW, HENRY J.: Pvt. Enl. on 4/19/61 at Martinsburg. Pres. on 8/31/61. In Farmville Gen. Hosp. 10/21-12/19/61 with debility. Discharged on 2/20/62, described as age 43, 5'11", dark comp., blue eyes, black hair, occ. laborer. Born in Frederick Co., MD.

TURNER, JAMES R.: Pvt. Enl. on 3/4/62 at Winchester. Pres. on 3/31/62 and 6/30/62 rolls.

VOGEL, JOHN A.: Pvt. Enl. on 4/19/61 at Martinsburg. Pres. on 8/31/61 as blacksmith. Absent in arrest at Centreville since 10/13/61 as Pvt. on 2/28/62 roll. Absent in arrest at Orange C.H. on 4/6/62. Transferred to Co. D, 2nd MD Cav. on 6/5/62 by S.O. 107. Deserted on 7/2/64.

WALKER, ELIAS M.: 1st Lt. Enl. on 4/19/61 at Martinsburg. Pres. on 8/31/61 as Pvt. Elected 1st Lt. on 2/10/62. Pres. on 2/28/62. Dropped at reorganization. Given tickets to return to Martinsburg, but found them useless, asked for alternate transportation to Gordonsville on 5/24/62.

WALKER, GEORGE WILLIAM: Pvt. B. 1843. Enl. on 1/31/62 at Centerville. Pres. on 2/28/62, 3/31/62 and 6/30/62 rolls. Transferred to Taylor's Battery on 10/8/62. Pres. on all rolls thru 2/28/65. D. 3/17/10, Asheville, N.C.

WANN, JOHN A.: 2nd Cpl. Enl. 5/18/61 at Harper's Ferry. Pres. on 8/31/61 roll. On 34 day furlough on 2/14/62. Pres. on 3/31/62 roll as pvt. Promoted to 2nd Cpl. by 6/30/62, when Pres. Transferred to Taylor's Battery on 10/8/62. Deserted on 7/17/63 at Bunker Hill. Taken POW at Martinsburg, WV on 7/26/63, released on oath of allegiance. Res. of Harford Co., MD. Enlisted in U.S. Marine Corps at Fort Mifflin, PA, described as 5'10", fair comp., black hair, blue eyes.

WARRING, JOHN L.: Postwar reference. Transferred to 1st Regt. Va. State Reserves.

WEDDEL, JOHN H., JR.: 2nd Sgt. Enl. on 4/19/61 at Martinsburg. Pres. on 8/31/61 and 2/28/62 rolls as Cpl. Promoted to Sgt. on 3/28/62. Pres. on 3/31/62 as 4th Sgt. and 6/30/62 as 2nd Sgt. rolls. Transferred to Taylor's Battery on 10/8/62. WIA in left leg at Chancellorsville, VA on 5/3/63. Medical certificate, Special Orders #271/19, assigned him as conscript officer for Caroline Co. Res. of Berkeley Co. Died on 9/8/1898, age 56 years 12 days. Buried in the Green Hill Cem., Martinsburg, WV.

WENTZ, HENRY D.: 1st Sgt. Enl. on 4/19/61 at Martinsburg. Pres. on 8/31/61 as Sgt. Pres. on 2/28/62 roll as 1st Sgt. Appointed 1st Sgt. on 2/10/62. Pres. on 3/31/62 and 6/30/62 rolls. Transferred to Taylor's Battery on 10/8/62. Pres. until taken POW at Harper's Farm, Amelia Co., on 4/6/65. Sent to Point Lookout, MD where held until released on oath on 6/21/65, res. of Berkeley Co., 5'8", dark comp., black hair, hazel eyes. B. Gettysburg, PA. Moved to Martinsburg in 1852. Occ. carriage maker. D. 12/10/1875, age 48. Buried in the Green Hill Cem., Martinsburg, WV.

WESTPHAL, CHARLES: Pvt. Enl. on 4/19/61 at Martinsburg. Deserted on 7/2/61. B. 6/21/1831. D. 2/14/1868. Buried in the Green Hill Cem., Martinsburg, WV.

WHITEHURST, JAMES: Cpl. Enl. on 8/14/61 at Manassas, transferred from the Danville Grays, 18th VA Inf. Pres. on 8/31/61 roll. On 35 day furlough on 2/14/62. Pres. on 3/31/62 and 6/30/62 rolls. Transferred to Taylor's Battery on 10/8/62. Pres. until KIA at Bean Station, TN in 1/64.

WILEY, WILLIAM A.: Postwar reference. B. 5/8/41. D. 11/11/06.

WIRT, HENRY D.: Sgt. Enl. on 4/19/61 at Martinsburg. Pres. on 8/31/61 roll as Pvt. Furloughed for 34 days on 2/14/62 as 4th Cpl. Pres. on 3/31/62 and 6/30/62 rolls as 3rd Cpl. Transferred to Taylor's Battery on 10/8/62. Taken POW on 7/19/63 at Martinsburg, Berkeley Co., WV, sent to Wheeling, WV, described as Res. of Martinsburg, age 22, 5'7", fair comp., auburn hair, blue eyes. Ordered released by U.S. Sec. of War on 9/9/63.

WRIGHT, JOHN P.: Pvt. Enl. on 3/31/62 at Orange C.H. Pres. on 6/30/62 roll. Transferred to Taylor's Battery on 10/8/62. Pres. most of the war. Paroled at Appomattox C.H., VA on 4/9/65. Res. of Orange Co.

WRIGHT, ROBERT J.: Pvt. Enl. on 3/31/62 at Orange C.H. Pres. on 3/31/62. AWOL in Orange Co. on 4/67/62. Pres. on 6/30/62.

WRIGHT, WILLIAM A.: Pvt. Enl. on 3/31/62 at Orange C.H. Pres. on 3/31/62. Sent to Gen. Hosp. #21 5/10/62, died there 6/3-4/62 of parotitis.

WRITTENHOUSE, P.: Postwar reference. D. 2/13/1905 near Mine Run. Buried in family cemetery.

YOUNG, JOHN R. B.: Blacksmith, Enl. on 4/19/61 at Martinsburg, WV. Pres. on 8/31/61, 2/28/62, 3/31/62, and 6/30/62 rolls. Transferred to Taylor's Battery on 10/8/62. Deserted 1/2/64 in TN. Died on 3/26/1883, buried in the Tomahawk Presbyterian Church Cem., Hedgesville, WV.

AILWORTH, EDMUND: Cpl. Enl. on 6/18/61 at Urbanna. Promoted to Cpl. on 4/28/62. Pres. on 4/30/62 roll. Discharged on 7/16/62, described as age 45, 5'7", dark comp., grey eyes, black hair, occ. farmer. Born in Accomack Co., VA. Res. of Urbanna, Middlesex Co.

ATKINSON, RICHARD (ROBERT) L.: Bugler. Enl. on 6/25/61 at Gloucester Point in Co. B, 55th VA Inf. (Or at Tappahannock, Essex Co.) Appointed Bugler on 4/30/62, appointed bugler the same day. Transferred from the Middlesex Art. on 10/4/62. Absent sick in hosp. on 2/28/63 roll. Never reported per 4/30/63 and later rolls. Discharged on 2/16/64 at Robertson Gen. Hosp. on 2/16/64. Described as age 37, 5'8", light comp., blue eyes, light hair, occ. merchant. Native Res. of Essex Co.

BEAZLEY, EDWARD W.: Pvt. Enl. on 5/27/61 at Urbanna in Co. B, 55th VA Inf. Sick at home 12/20/61-2/62. Hospitalized on 4/28/62. Transferred to Jackson Flying Art. on 10/4/62. Pres. on 2/28/63, 4/30/63 as Pvt. In Farmville Gen. Hosp. 6/10-7/1/63 with scabies. Promoted to 8th Cpl. on 8/10/63. Pres. on 8/31/63 and 10/31/63 rolls. Absent on furlough on 12/31/63. Pres. on 2/29/64 as 8th Cpl. In Chimborazo Hosp. #1 with acute diarrhea, returned to duty on 7/5/64. Pres. on 10/31/64, 12/31/64, and 2/28/65 rolls as 2nd Cpl. Res. of Middlesex Co.

BEAZLEY, GEORGE PHILIP: Sgt. Enl. on 5/24/61 at Urbanna in Co. B, 55th VA Inf. Promoted to Sgt. on 4/29/62. Pres. on 4/30/62. Later served as a pvt. in Clutter's Battery. Transferred to Co. G, 9th VA Cav. on 3/15/63. WIA at White's Tavern on 8/16/64. Pres. thru at least 10/1/64. Transferred to Jackson Flying Art. on 10/4/62. Paid $50 bounty on 10/31/62. In Gen. Hosp. #20, Richmond on 12/20/62. In Camp Winder Hosp. on 12/21/62. Absent sick in hosp. on 2/28/63. Pres. on 4/30/63. Sent to Hosp. from Hamilton Crossing on 6/6/63 per 8/31/63 roll. Furloughed for 30 days on 7/18/63 from Farmville Gen. Hosp., had been in the hosp. for "spermatic cord atrophy of testicle following ochitits, feebleness and general debility. Did not return to Jackson Flying Art. Transferred to Co. G, 9th VA Cav. on 3/15/63 by GS.O. 74, para 1, HQ, Army of Northern Virginia. WIA at White's Tavern on 8/16/64. Pres. thru at least 10/1/64. Paroled at Mechanicsville, VA on 4/20/65. B. Saluda, Middlesex Co. on 4/7/1840. Res. of Caroline Co. during the war. D. Saluda on 12/19/1917.

BLACKLEY, WILLIAM H.: Pvt. Enl. on 12/1/61 in Co. B, 55th VA Inf. Went home sick on 3/1/62. Pres. sick on 4/30/62. Discharged on 6/20/62, described as age 18, 5'7", born in Essex Co., occ. farmer, light comp., blue eyes, light hair.

BRAY, GEORGE W.: Pvt. Enl. on 3/17/62 at Port Lowry. Previously served in Co. A, 30th VA Inf, enlisting there on 4/22/61, then age 22. Pres., sick, on 4/30/62. Later served in Kirkpatrick's' Battery. Taken POW at Waynesboro, VA on 3/2/65, sent to Fort Delaware where he was held until the end of the war. Res. of Middlesex Co. and Fredericksburg, VA.

BRAY, RICHARD B.: Pvt. Enl. on 8/10/61 in Co. B, 55th VA Inf. Detailed as carpenter with regimental QM in 1/62. Pres. on 4/30/62. Discharged on 7/16/62, described as age 48, 6', dark comp., blue eyes, gray hair, occ. farmer. Born in Middlesex Co.

BREEDEN, JAMES H.: Pvt. Enl. in Co. H, 3rd VA Art. LDT. Transferred to Capt. Fleet's Battery (Middlesex Art.) on 5/14/62 by S.O. 21, HQ, 1st Brigade. Discharged on 7/28/62.

BREEDEN, NATHAN: Pvt. Enl. in Co. H, 3rd VA Art. LDT. Transferred to Capt. Fleet's Battery (Middlesex Art.) on 5/14/62 by S.O. 21, HQ, 1st Brigade. Later served in the Amherst Art. Res. of Rockingham Co., VA postwar.

BREEDEN, WILLIAM: Pvt. Enl. in Co. H, 3rd VA Art. LDT. Transferred to Capt. Fleet's Battery (Middlesex Art.) on 5/14/62 by S.O. 21, HQ, 1st Brigade. Later served in the Amherst Art. Served until taken POW at Waynesboro, VA on 3/2/65, sent to Fort Delaware, where held until released on oath on 6/19/65. Described as res. of Henrico Co., ruddy comp., light hair, blue eyes, 5'8".

BRISTOW, CALVIN: Pvt. Enl. on 5/24/61 at Urbanna (Gloucester Point) in Co. B, 55th VA Inf. Pres. on 4/30/62. Reassigned to the Ashland Art. on 10/4/62 by S.O. 209, Headquarters, Army of Northern Virginia. May have been WIA at Gettysburg, PA. Pres. on all rolls thru 2/28/65. Paroled at Appomattox C.H., VA on 4/9/65. B. 3/22/1841 in Middlesex Co. Living in Saluda Twp., Middlesex Co. farmer, in 1870. D. 7/6/1931 in Middlesex Co.

BRISTOW, FRANCIS A.: Pvt. Enl. on 6/27/61 at Gloucester Point in the 26th VA Inf. Transferred to Co. B, 55th VA Inf. on 11/12/61. Pres. on 4/30/62 roll. Reassigned to the Ashland Art. on 10/4/62 by S.O. 209, Headquarters, Army of Northern Virginia. Hospitalized some parts of the war. Paroled at Appomattox C.H., VA on 4/9/65. Living in Saluda Twp., Middlesex Co., VA, age 28 dry goods merchant in 1870.

BRISTOW, GEORGE T.: Pvt. Enl. on 5/24/61 at Urbanna, VA. Pres. sick on 4/30/62. Discharged on 7/16/62, described as age 35, 5'11", light comp., gray eyes, light hair, occ. farmer. born in Middlesex Co.

BRISTOW, HENRY: Pvt. Enl. on 5/24/61 at Urbanna, VA in Co. B, 55th VA Inf. Detailed to guard provisions 10/28/61-11/61. Employed as carpenter 1/1-11/62. Absent sick in Hosp. on 4/28/62. In Chimborazo Hosp. #2 for chronic dysentery 8/17-31/62. Furloughed for 15 days on 8/31/62. Transferred to Jackson Flying Art. on 10/4/62. Absent sick in Hosp. on 2/28/63 roll. Never reported to Jackson Flying Art.

BRISTOW, JAMES BENJAMIN: Pvt. Enl. on 5/24/61 at Urbanna, VA in Co. B, 55th VA Inf. Discharged on 12/31/61 upon furnishing a substitute. May have later served in Co. C, 26th VA Inf.

BRISTOW, LEMUEL C.: Sr. 2nd Lt. Enl. on 5/24/61 at Urbanna, VA in Co. B, 55th VA Inf. Promoted to Sr. 2nd Lt. on 4/28/62. Pres. on 4/30/62. Later served as enrolling officer in Gloucester Co. Paroled as a member of Fleet's Battery at Burkeville, VA on 4/26/65.

BRISTOW, WILLIAM C.: Sgt. Enl. on 5/24/61 at Urbanna, VA in Co. B, 55th VA Inf. Pres. sick on 4/30/62. Transferred to Jackson Flying Art. on 10/4/62. Pres. on 2/28/63 thru 2/29/64 rolls as Pvt. Pres. on 10/31/64, 12/31/64, and 2/28/65 rolls as 7th Cpl.

BROWN, JOHN H.: Pvt. Enl. on 2/26/62 at Urbanna. Transferred to Jackson Flying Art. on 10/4/62. Pres. thru 12/31/63 roll. Absent on furlough on 2/29/64. In Farmville Gen. Hosp. 3/21/64. Furloughed for 30 days on 3/22/64. In Wayside Hosp. 4/18-19/64. In Farmville Gen. Hosp. 3/22-4/19/64 with chronic bronchitis. Pres. on 10/31/64, 12/31/64 and 2/28/65 rolls. Paroled at Appomattox C.H., VA on 4/9/65. Res. of Cobham, Albemarle Co., VA.

BURCH, CHARLES: Pvt. Enl. on 3/1/62 in Urbanna, VA in Co. B, 55th VA Inf. Pres. sick on 4/30/62. Transferred to Jackson Flying Art. on 10/4/62. Severely WIA at Fredericksburg, VA on 12/13/62. Did not return to unit. Retired to Invalid Corps P.A.C.S. on 6/16/64.

BURROWS, MASON M.: Musician. Enl. on 5/24/61 at Urbanna in Co. B, 55th VA Inf. Elected musician on 7/19/61. Appointed chief musician of the 55th VA Inf. on 12/4/61. Transferred to C.S. Navy on 4/17/62, appointed Acting Master Mate. Joined Major Morris' Signal Corps as signal sgt. in 10/62, served in Richmond, VA and Charleston, SC. Resigned on 11/2/63, resignation forwarded with much reluctance.

CARLTON, THOMAS F.: Pvt. Enl. on 5/24/61 at Urbanna in Co. B, 55th VA Inf. Sick at home 2-3/62. Absent sick in hosp. since 4/27/62 per 4/30/62 roll. Died in service. John G. Anderson, attorney, filed claim for pay and allowances on 11/24/62.

CARTER, EDWARD: Pvt. Enl. on 8/26/62 in Loudoun Co. by Capt. Fleet. Transferred to Jackson Flying Art. on 10/4/62. Absent on sick furlough on 2/28/63. Sent to Hosp. from Bunker Hill in 11/62, never rejoined.

CHANDLER, JOHN J.: 2nd Sgt. Pvt. Enl. on 8/7/61 at Church View, VA, in Co. B, 55th VA Inf. Pres. on 4/30/62. Transferred to Jackson Flying Art. on 10/4/62. Pres. on 2/28/63 and 4/30/63 as Pvt. Promoted to 4th Cpl. on 8/10/63. Pres. on 8/31/63, 10/31/63, and 12/31/63 as 4th Cpl. Pres. on 2/29/64 as 2nd Cpl. Pres. on 10/31/64 as 2nd Sgt. Pres. on 12/31/64 and 2/28/65 as 3rd Sgt. Paroled on 5/2/65.

CHANDLER, SILAS: Pvt. Enl. on 5/24/61 (6/15/61) at Urbanna in Co. B, 55th VA Inf. Left to guard provisions at Camp Ashby 10/28/61-11/8/61. Detailed to camp guard at Camp Ashby 4/30/62 roll. WIA and taken POW at Sharpsburg, MD on 9/30/62, minnie ball in the left knee. Paroled at Fort McHenry, sent to Fort Monroe for exchange. Admitted to Chimborazo Hosp. 10/24/62 for leg wound. Reassigned to the Ashland Art. on 10/4/62 by S.O. 209, Headquarters, Army of Northern Virginia. WIA at Gettysburg, PA, canister ball to the shoulder. Promoted to 5th Cpl. Pres. thru 2/28/65. Paroled at Appomattox C.H., VA on 4/9/65. B. York Co., VA. Letters to his wife exist. Living in Saluda Twp, Middlesex Co., VA in 1870 then a 38 year old farmer. Applied in 1906 for a pension then age 75. D. 4/1912 at Urbanna, Middlesex Co., VA, age 81.

CHOWNING, JOHN: Pvt. Enl. on 5/24/61 at Urbanna in Co. B, 55th VA Inf. Went home sick 2/25/62. Absent on 3/31/63. Pres. sick on 4/30/62. Paid for service thru 8/31/62. Described as age 19, gray eyes, light hair, dark comp., 5'11", born in Middlesex Co., VA., occ. farmer. Transferred to Jackson Flying Art. on 10/4/62. Pres. on 2/28/63 thru 12/31/63 rolls. Absent detailed with wagon train on 2/29/64. WIA in right shoulder. Admitted to Chimborazo Hosp. #3 on 8/8/64 with gunshot wound in left shoulder. Absent on sick furlough on 10/31/64. Admitted to Chimborazo Hosp. #2 on 12/23/64. Transferred to Chimborazo #4 on 12/24/64. Absent sick on 12/31/64. Returned to duty on 2/17/65. Pres. on 2/28/65. Paroled at Appomattox C.H., VA on 4/9/65.

CHOWNING, WILLIAM B.: Pvt. Enl. on 5/24/61 at Urbanna in Co. B, 55th VA Inf. Pres. sick on 4/30/62. Admitted to Gen. Hosp. #22 on 10/18/62. Transferred to Jackson Flying Art. on 10/4/62. Pres. on 2/28/63 and 4/30/63 rolls. Taken POW on 6/29/63 in PA per 8/31/63 and C.S.A. rolls thru 10/31/64. Taken POW on 6/28/63 at Williamsburg, PA per Federals, sent to Fort Delaware, where held until exchanged on 4/27/64. WIA. Admitted to Chimborazo Hosp. #1 on 8/15/64 for wounds. Absent, detailed in Richmond Hosp. per 12/31/64 and 2/28/65 rolls, declared unfit for field duty. Returned to duty on 3/25/65. Paroled at Appomattox C.H., VA on 4/9/65.

CLARK, JOHN C.: Pvt. Enl. on 5/27/61 at Urbanna, VA in Co. B, 55th VA Inf. Absent in hosp. on 4/30/62. Transferred to Jackson Flying Art. on 10/4/62. Pres. on 2/28/63 and 4/30/63 rolls as Pvt. Promoted to 6th Cpl. on 8/10/63. Pres. on 8/31/63, 10/31/63, 12/31/63, and 2/29/64 as 6th Cpl. Pres. on 10/31/64, 12/31/64 and 2/28/65 rolls as 3rd Cpl.

CLAYBROOK, WILLIAM LOUIS.: Sgt. Graduated 1847 from Columbian College, Washington, D.C. Enl. on 5/24/61 at Urbanna, VA. Promoted to Sgt. from Cpl. on 4/28/62. Pres. on 4/30/62.

COGHILL, ARCHIBALD: Pvt. Enl. on 7/15/61 at Camp Sullivan. Pres. sick on 4/30/62. Transferred to Jackson Flying Art. on 10/4/62. Pres. on 2/28/63 and all rolls thru 2/28/65. Paroled at Ashland, VA on 6/1/65. Occ. farmer.

CONNER, JOHN R.: Pvt. Enl. in Co. H, 3rd VA Art. LDT. Transferred to Capt. Fleet's Battery (Middlesex Art.) on 5/14/62 by S.O. 21, HQ, 1st Brigade.

COVINGTON, JOSEPH GREGORY: Pvt. B. Essex Co., Va. Enl. on 5/24/61 at Urbanna, VA in Co. B, 55th VA Inf. Detailed as a carpenter 1-2/62. Pres. on 4/30/62. Admitted to Gen. Hosp., Staunton, VA on 8/9/62. Transferred to Jackson Flying Art. on 10/4/62. Pres. on 2/28/63 and 4/30/63 rolls. Sent to Hosp. from Culpeper C.H. on 8/1/63. Pres. on 10/31/63 and all subsequent rolls. In Gen. Hosp. #9 on 6/28/64, furloughed for 30 days on 6/29/64. Paroled at Appomattox C.H., VA on 4/9/65. D. 1/7/06, Middlesex Co., Va., age 68.

CRAFTON, ANDREW N.: Pvt. Enl. on 5/24/61 at Urbanna, VA in Co. B, 55th VA Inf. Pres. on 4/30/62. Transferred to Jackson Flying Art. on 10/4/62. Pres. on 2/28/63 and all later rolls. Paroled at Appomattox C.H., VA on 4/9/65.

CRAFTON, ROBERT F.: Pvt. Enl. on 5/24/61 at Urbanna, VA in Co. B, 55th VA Inf. Pres. sick on 4/30/62. Transferred to Jackson Flying Art. on 10/4/62. Pres. on 2/28/63 and all subsequent rolls thru 2/28/65. In Chimborazo Hosp. #1 on 3/20/65. RTD on 3/21/65.

CRITTENDEN, BERRY: Pvt. Enl. on 3/4/62 in Richmond by Capt. Carrington, Transferred from Co. H, 3rd VA Art. LDT on 5/14/62 by S.O. 21, HQ, 1st Brigade. Transferred to Jackson Flying Art. on 10/4/62. Joined on 11/25/63. Pres. on 12/31/63 and 2/29/64 rolls. Taken POW on 10/7/64 per 10/31/64 and later Confederate rolls. No Federal record.

DUNN, THOMAS H.: Drummer. Enl. on 6/12/61 at Urbanna, VA in Co. B, 55th VA Inf. Transferred to Co. K, 55th VA Inf. on 5/1/62. Pres. on all rolls thru 12/31/64. Entered R. E. Lee Camp Soldiers Home on 4/11/1906, died there on 8/2/1911, age 78. Buried in the Hollywood Cem., Richmond, VA.

DURHAM, J. J.: Pvt. Enl. in Co. I, 3rd VA Art. LDT. Transferred to Capt. Fleet's Battery (Middlesex Art.) on 5/14/62 by S.O. 21, HQ, 1st Brigade.

ENROUGHTY, H.: Pvt. Enl. in Co. H, 3rd VA Art. LDT. Transferred to Capt. Fleet's Battery (Middlesex Art.) on 5/14/62 by S.O. 21, HQ, 1st Brigade.

ENROUGHTY, HARRISON: Pvt. Enl. in Co. H, 3rd VA Art. LDT. Transferred to Capt. Fleet's Battery (Middlesex Art.) on 5/14/62 by S.O. 21, HQ, 1st Brigade.

EVANS, ZACHALINE M.: Pvt. Enl. in Co. H, 3rd VA Art. LDT. Transferred to Capt. Fleet's Battery (Middlesex Art.) on 5/14/62 by S.O. 21, HQ, 1st Brigade.

FARMER, LAWSON F.: Pvt. Enl. on 7/2/61 at Urbanna, VA in Co. B, 55th VA Inf. Pres. on 4/30/62. Later served in Marmaduke Johnson's (Jackson Flying Art.) Battery. Absent sick in hosp. on 2/28/63 roll since 12/20/62. AWOL from Hosp. from 3/18/63 to 8/28/63. Pres. on 8/31/63 thru 10/31/64 rolls. Absent on furlough on 12/31/64. Pres. on 2/28/65. Paroled at Appomattox C.H., VA on 4/9/65.

FIGG, JOHN H.: Pvt. Enl. on 5/27/61 at Urbanna. On 1/17/62 requested detail as a wheelwright to work on making wagons "or any thing of the kind." Detailed to work on gun boats by Sec. of War on 2/27/62 per 4/30/62 roll. Discharged on 6/30/62, described as age 25, 5'7", light comp., blue eyes, light hair, occ. wheelwright, born in Gloucester Co. Buried in the Ebenezer Baptist Church Cem., Gloucester Co.

FILCHETT, SALATHIEL P.: Pvt. Enl. on 5/24/61 at Urbanna, VA in Co. B, 55th VA Inf. Absent sick at home since 4/18/62 per 4/30/62 roll. Reassigned to the Ashland Art. on 10/4/62 by S.O. 209, Headquarters, Army of Northern Virginia. Promoted to Cpl. and Sgt. D. 6/1892 per widow's pension application. Res. of Middlesex Co.

FISHER, RICHARD L.: Pvt. Enl. on 7/7/61 at Urbanna, VA in Co. B, 55th VA Inf. Pres. sick on 4/30/62. Res. of Essex Co., VA.

FLEET, WILLIAM C.: Captain. Enl. on 5/24/61 at Urbanna, VA in Co. B, 55th VA Inf. Reelected Capt. on 4/28/62. Pres. on 4/30/62. Not Pres. during summer of 1862. Under arrest in 9/62 by order of Gen. Jackson. Later served as enrolling officer, 1st Congressional District, VA (which was composed of

Middlesex, Accomack, Northampton, King William, Gloucester, Mathews, Lancaster, Westmoreland, Richmond, Essex, King & Queen, and Northumberland counties). D. Palestine, Texas, age 74 about 5/1/1910.

FORE, JOHN A.: Pvt. Enl. in Co. I, 3rd VA Art. LDT. Transferred to Capt. Fleet's Battery (Middlesex Art.) on 5/14/62 by S.O. 21, HQ, 1st Brigade.

GARDNER, ALEXANDER SAMUEL: Drummer. Enl. on 5/24/61 at Urbanna, VA in Co. B, 55th VA Inf. Appointed drummer on 1/1/62. Absent sick in Hosp. since 4/18/62 per 4/30/62 roll. Later served in the Amherst Art. He was KIA at Monocacy Creek on 7/9/64, buried in the Mt. Olivet Cem., Frederick, MD.

GARDNER, THOMAS: Pvt. Enl. on 7/24/61 at Urbanna, VA in Co. B, 55th VA Inf. Detailed as a carpenter 1/1/-31/62. Pres. on 4/30/62. Transferred to Jackson Flying Art. on 10/4/62. Absent sick in Hosp. on 2/28/63. Pres. on 4/30/63 and all rolls thru 12/31/64. In Chimborazo Hosp. #3 on 11/16/64 with intermittent fever. Returned to duty on 11/29/64. Absent, detailed in QM Dept., Johnson's Battn. on 2/28/65. Paroled on 5/19/65.

GARRETT, JAMES R.: Pvt. Enl. on 5/24/61 at Urbanna, VA in Co. B, 55th VA Inf. Pres. on 4/30/62. Transferred to Jackson Flying Art. on 10/4/62. Pres. on 2/28/63. Absent sick on 4/30/63. Admitted to Chimborazo Hosp. #1 on 4/17/63 with pneumonia. Died in Richmond on 4/21/63.

GAYLE, JOHN Z.: Pvt. Enl. on 5/24/61 at Urbanna, VA in Co. B, 55th VA Inf. Absent sick on 3/9- 31/62. Pres. on 4/30/62 roll. Reassigned to the Ashland Art. on 10/4/62 by S.O. 209, Headquarters, Army of Northern Virginia. WIA in right side above hip at Gettysburg, PA on 7/2/63. Taken POW at Cashtown, PA on 7/5/63, sent to Gen. Hosp. at Chester, PA, and then to Point Lookout, MD. Exchanged. Paroled at Appomattox C.H., VA on 4/9/65. Living Saluda Twp, Middlesex Co., age 30 in 1870.

GAYLE, WILLIAM A. (J.?): Pvt. Enl. on 5/24/61 at Urbanna, VA in Co. B, 55th VA Inf. Pres. on 4/30/62 roll. Reassigned to the Ashland Art. on 10/4/62 by S.O. 209, Headquarters, Army of Northern Virginia. Detailed to wait on the wounded at Gettysburg, PA. Taken POW at Cashtown, PA on 7/5/63, sent to Point Lookout, MD. Exchanged at City Point, VA on 3/16/64. Paroled at Appomattox C.H., VA on 4/9/65. Living in Saluda Dist., Middlesex Co. in 1870 then a 30-yr old farmer. Bro. of John Z. Gayle.

GILLS, GEORGE: Pvt. Enl. on 8/24/61 at Urbanna, VA in Co. B, 55th VA Inf. Reassigned to the Ashland Art. on 10/4/62 by S.O. 209, Headquarters, Army of Northern Virginia. Pres. 10/31/62. Absent sick 12/31/62. In Gen. Hosp. #9 2/28/63 roll. Pres. remaining rolls thru 2/28/65.

GILLS, THOMAS JEFFERSON: Pvt. Enl. on 2/12/62 at Urbanna, VA in Co. B, 55th VA Inf. Reassigned to the Ashland Art. on 10/4/62 by S.O. 209, Headquarters, Army of Northern Virginia. Pres. on all rolls, but Taken POW at Gettysburg, PA, but not listed in Federal POW records. Paroled at Appomattox C.H., VA on 4/9/65. Res. of Amelia Co.

GOODE, GEORGE F.: Pvt. Enl. on 5/24/61 at Urbanna, VA in Co. B, 55th VA Inf. Sick at home 12/21/61-3/62. Pres. on 4/30/62. Later served in the Amherst Art. Taken POW at Waynesboro, VA on 3/2/65, sent to Ft. Delaware where held until released on 6/1/65. Described as 5'8", ruddy comp., dark hair, blue eyes. Res. of Middlesex Co.

GOODE, JOHN C.: Pvt. Enl. on 3/1/62 in Co. B, 55th VA Inf. Previously served in Co. B, 109th VA Militia. Pres. on 4/30/62. Transferred to Jackson Flying Art. on 10/4/62. Pres. on 2/28/63 and all later rolls. Admitted to Gen. Hosp. 10/9/63. Transferred to Chimborazo on 10/10/63 with chronic rheumatism. Returned to duty on 10/17/63. Paroled at Mechanicsville, VA on 5/5/65. He is also listed on PWR of the Amherst Art. Res. of Middlesex Co., VA.

GOODE, SYLVANUS: Pvt. Enl. in Co. H, 3rd VA Art. LDT. Transferred to Capt. Fleet's Battery (Middlesex Art.) on 5/14/62 by S.O. 21, HQ, 1st Brigade. Later served in the Amherst Art., was taken POW at Waynesboro, VA, sent to Fort Delaware, where held until he died on 7/3/65 of pneumonia. Buried in the Finn's Point Cem., Salem, NJ.

GOODE, WILLIAM H.: Pvt. Enl. on 12/31/61 at Urbanna, VA. Pres. on 4/30/62. Later served in the Amherst Art., and served until MWIA in the thigh on 7/20/64 at Stephenson's Dept, and died on 7/28/64 and was "buried by the good people of Winchester."

HALL, ROBERT: Pvt. Enl. on 5/24/61 at Urbanna, VA in Co. B, 55th VA Inf. Pres. on 4/30/62. In Chimborazo Hosp. #3 8/17-9/18/62 with debilities. Transferred to Jackson Flying Art. on 10/4/62. Pres. on 2/28/63 thru 10/31/63 rolls. In Gen. Hosp. #9 12/25-26/63. In Gen. Hosp. #9 1/6-7/64. Absent on furlough on 12/31/64. Pres. on 2/29/64 and all later rolls. Paroled in Richmond on 5/20/65.

HARDY, ANDREW STRAUGHN: Pvt. Served in this unit according to documents on file with the Amherst Art. Taken POW at Waynesboro, VA on 3/2/65, sent to Fort Delaware. Released on 6/20/65, described as Res. of Middlesex Co., light comp., dark hair, brown eyes, 5'11". Moved to Lancaster Co., VA postwar, and to NY ca. 1902. Died on 1/12/1929 in Johnson City, TN, buried in White Stone Baptist Church Cem., Lancaster Co., VA. Bro. of John T.

HARDY, JOHN T.: Pvt. Enl. on 5/24/61 at Urbanna, VA in Co. B, 55th VA Inf. Pres. on 4/30/62. Later served in Kirkpatrick's Battery (Amherst Art.) for the rest of the war. Res. of Middlesex Co. B. 6/3/1842 in Middlesex Co. Farmer postwar. D. 4/5/1918 (1916). Bro. of William B. Hardy.

HARDY, WILLIAM BENJAMIN: 1st Lt. VMI, 63'.Enl. on 11/23/61 at Urbanna, VA in Co. B, 55th VA Inf. Promoted to 1st Lt. on 4/28/62. Commanded the Middlesex Art. during the summer of 1862, particularly prominent in battle of Cedar Mountain. In 9/62, "had celebrated stand up row with General Pendleton, which led to his arrest. He later refused to apologize and was relieved from duty on 10/4/62 to await court-martial." Appointed enrolling officer for Essex Co. on 12/22/62. Taken POW by Federal raiders on 5/22/63 in Middlesex Co., VA, sent to Johnson's Island, OH where held until exchanged on 2/20/65. D. 1913.

HARMAN, LEMOCUM: Pvt. Enl. in Co. H, 3rd VA Art. LDT. Transferred to Capt. Fleet's Battery (Middlesex Art.) on 5/14/62 by S.O. 21, HQ, 1st Brigade.

HART, WILLIAM H.: Pvt. Enl. on 5/24/61 at Urbanna, VA in Co. B, 55th VA Inf. Pres. on 4/30/62 roll. Reassigned to the Ashland Art. on 10/4/62 by S.O. 209, Headquarters, Army of Northern Virginia. Pres. thru 2/28/65. Slightly WIA in arm at Gettysburg, PA. Paroled at Appomattox C.H., VA on 4/9/65.

HEALY, JAMES T.: Pvt. Enl. on 5/24/61 at Urbanna. Pres. on 4/30/62. Transferred to Jackson Flying Art. on 10/4/62. Never reported per 2/28/63 roll. Pres. on 4/30/63. Died at Guinea Hosp. on 6/6/63.

HEALY, WILLIAM H.: Pvt. Enl. on 5/24/61 at Urbanna in Co. B, 55th VA Inf. Pres. on 4/30/62. Reassigned to the Ashland Art. on 10/4/62 by S.O. 209, Headquarters, Army of Northern Virginia. Pres. on all rolls thru 2/28/65. D. 8/1904, buried in family Cem. in West Point, New Kent Co., VA.

HEATH, GEORGE HENRY: Pvt. Enl. on 5/24/61 at Urbanna in Co. B, 55th VA Inf. Absent in Hosp. 4/30/62 roll since 4/20/62. Admitted to Soldier's Home Hosp., Richmond, on 8/1/62. Transferred to Lynchburg Hosp. 8/8/62. Reassigned to the Ashland Art. on 10/4/62 by S.O. 209, Headquarters, Army of Northern Virginia. Pres. 10/31/62. Absent sick in Chimborazo Hosp. #1 12/21/62 with pneumonia. RTD 3/3/63. Pres. on all rolls thru 2/28/65. B. Gloucester C.H. on 6/9/1836. Living in Middlesex Co., VA in 1870. D. at Urbanna on 3/3/1904.

HEATH, JOSEPH S.: Pvt. Enl. on 6/18/61 at Urbanna, VA in Co. B, 55th VA Inf. Absent sick in Hosp. since 4/25/62 per 4/30/62 roll. Transferred to Jackson Flying Art. on 10/4/62. Pres. on 2/28/63 and all later rolls. Paroled at Appomattox C.H., VA on 4/9/65.

HILLIARD, RICHARD D.: Pvt. Enl. on 5/24/61 at Urbanna, VA in Co. B, 55th VA Inf. Pres. sick on 4/30/62. Reassigned to the Ashland Art. on 10/4/62 by S.O. 209, Headquarters, Army of Northern Virginia. Pres. on all rolls thru 12/31/64. Absent on furlough on 2/28/65. D. 1917.

HOWLETT, JAMES P.: Pvt. Enl. on 5/24/61 at Urbanna, VA. Absent in hosp. since 4/28/62 per 4/30/62 roll. Transferred to Jackson Flying Art. on 10/4/62. Pres. on 4/30/63. Absent on 8/31/63 roll, had been MWIA on 8/1/63 near Culpeper C.H. Died on 8/22/63 at Charity Hosp., Gordonsville, VA.

HUNDLEY, RICHARD H.: Pvt. Enl. on 3/9/62 at Gloucester Point, $50 bounty due. Pres. on 4/30/62. Transferred to Jackson Flying Art. on 10/4/62. Pres. on 2/28/63 and 4/30/63 rolls. KIA at Gettysburg, PA on 7/2/63.

HUTCHINSON, HENRY: Pvt. Enl. in Co. I, 3rd VA Art. LDT. Transferred to Capt. Fleet's Battery (Middlesex Art.) on 5/14/62 by S.O. 21, HQ, 1st Brigade.

INGE, WILLIAM P.: Pvt. Enl. in Co. I, 3rd VA Art. LDT. Transferred to Capt. Fleet's Battery (Middlesex Art.) on 5/14/62 by S.O. 21, HQ, 1st Brigade.

ISOM, WILLIAM W.: Pvt. Enl. on 5/24/61 at Urbanna, VA in Co. B, 55th VA Inf. Pres. on 4/30/62. Transferred to Jackson Flying Art. on 10/4/62. Pres. on 2/28/63 and all rolls thru 2/28/65. Taken POW in last days of the war near Fredericksburg. Sent to Point Lookout, MD where held until released on oath on 6/14/65. Described as 5'6-", fair comp., auburn hair, blue eyes. Res. of King and Queen Co.

JOHNSON, WILLIAM: Pvt. Enl. in Co. H, 3rd VA Art. LDT. Transferred to Capt. Fleet's Battery (Middlesex Art.) on 5/14/62 by S.O. 21, HQ, 1st Brigade.

KEINGINGHAM, GIDEON: Cpl. Enl. on 5/24/61 at Urbanna in Co. B, 55th VA Inf. Detailed as carpenter for regimental carpenter 12/31/61. Promoted to Cpl. on 4/28/62. Pres. sick on 4/30/62 roll. Reassigned to the Ashland Art. on 10/4/62 by S.O. 209, Headquarters, Army of Northern Virginia. Pres. on all subsequent rolls thru 2/28/65. WIA in face at Gettysburg, PA.

KEININGHAM, JAMES: 2nd Lt. Enl. on 5/24/61 at Urbanna, VA in Co. B, 55th VA Inf. Resigned on 11/5/61.

LAWSON, WILLIAM J.: Pvt. Enl. on 5/24/61 at Urbanna, VA in Co. B, 55th VA Inf. Pres. on 4/30/62. Transferred to Jackson Flying Art. on 10/4/62. Absent WIA at Warrenton Springs on 2/28/63 roll. NFR.

LEWIS, J. P.: Pvt. Enl. on 10/26/61 at Urbanna, VA. Transferred to Jackson Flying Art. on 10/4/62. Died of pneumonia on 11/14/62 in Chimborazo Hosp. #3.

LEWIS, JOHN T.: Pvt. Enl. on 10/26/61 at Camp Harmony, Urbana, VA. Pres. sick on 4/30/62. Died of typhoid pneumonia at Chimborazo Hosp. #3 on 11/14/62.

LUCKADOE, THOMAS S.: Pvt. Enl. on 2/27/62 at Urbanna. Transferred to Jackson Flying Art. on 10/4/62. Died in service. Claim for pay and allowances filed on 9/9/63 by E. Luckadoe, father.

MAJOR, LEONARD OSMAN BRISTOW: Pvt. Enl. on 5/24/61 at Urbanna, VA in Co. B, 55th VA Inf. Pres. on 4/30/62. Pres. on 2/28/63 and subsequent rolls thru 2/29/64. WIA in leg and forearm on 5/11/64 at Spotsylvania C.H., sent to Hosp. Arm amputated at Charlottesville Gen. Hosp. on 5/19/64, permanently disabled. Retired to Invalid Corps. P.A.C.S. on 1/18/65. D. 2/18, Middlesex Co., Va., age 76.

MASSEY, JOHN R.: Pvt. Enl. on 5/24/61 at Urbanna, VA in Co. B, 55th VA Inf. Pres. sick on 4/30/62. Transferred to Jackson Flying Art. on 10/4/62. Pres. on 2/28/63 and 4/30/63 rolls. MWIA near Culpeper C.H., VA, sent to Hosp. at Gordonsville, VA where he died on 8/10/63 or wounds or Erysipelas.

McKANN, PHILIP H.: Pvt. Enl. on 5/24/61 at Urbanna, VA in Co. B, 55th VA Inf. Pres. on 4/30/62. Transferred to Jackson Flying Art. on 10/4/62. Pres. on 2/28/63, 4/30/63, and 6/30/63 rolls. Sent to Hosp. from Rapidan Station, VA on 10/8/63 per 10/31/63 roll. In Chimborazo Hosp. #5 10/11-12/21/63 with chronic diarrhea. Pres. on 12/31/63. Absent on furlough on 2/29/64. Pres. on 10/31/64, 12/31/64, and 2/28/65 rolls. Paroled at Appomattox C.H., VA on 4/9/65. D. 2/10/01 in Middlesex Co., age 69.

McKINLEY, DANIEL: Pvt. Enl. in Co. D, 3rd VA Art. LDT. Transferred to Capt. Fleet's Battery (Middlesex Art.) on 5/14/62 by S.O. 21, HQ, 1st Brigade.

MINSON, JAMES W.: Pvt. Enl. in Co. H, 3rd VA Art. LDT. Transferred to Capt. Fleet's Battery (Middlesex Art.) on 5/14/62 by S.O. 21, HQ, 1st Brigade. Later served in the Amherst Art. Described as occ. farmer, dark comp., dark hair, gray eyes, 5'6".

MONTAGUE, CHRISTMAS D.: 1st Sgt. Enl. on 5/24/61 at Urbanna, VA in Co. B, 55th VA Inf. Promoted to 1st Sgt. from Sgt. on 4/28/62. Absent sick at Ashland Hosp. since 4/16/62 per 4/30/62 roll. Transferred to Jackson Flying Art. on 10/4/62. Pres. on 10/31/64, 12/31/64, and 2/28/65 rolls as Pvt. Paroled at Appomattox C.H., VA on 4/9/65 as 2nd Lt.

MONTAGUE, THOMAS H.: Pvt. Enl. on 3/1/62 in Co. B, 55th VA Inf. Previously served in Co. B, 109th VA Militia. Absent sick in hosp. since 4/20/62 per 4/30/62 roll. Transferred to Jackson Flying Art. on 10/4/62. First record in Jackson's Flying Art. Pres. on 2/29/64 roll. Pres. on 10/31/64. Absent on furlough on 12/31/64. Pres. on 2/28/65. Paroled at Appomattox C.H., VA on 4/9/65.

MUSE, WILLIAM H.: Pvt. Enl. on 5/24/61 at Urbanna, VA in Co. B, 55th VA Inf. Absent in Middlesex Co. sick since 3/25/62 per 4/30/62 roll. In Gen. Hosp. #10 10/24-27/62. Transferred to Jackson Flying Art. on 10/4/62. Absent sick on 2/28/63 roll. Admitted to Gen. Hosp. #12 on 1/21/63 for fracture of tibia of both legs by cannon wheel producing chronic ulcers and retraction of tendons. Discharged on 1/22/63. Described as age 33, 6'2", dark comp., dark eyes, gray hair. Born in Urbanna, Middlesex Co., VA.

NEWBILL, JOHN H.: Pvt. Enl. on 5/24/61 (12/1/61) at Urbanna, VA in Co. B, 55th VA Inf. Pres. on 4/30/62. Transferred to Jackson Flying Art. on 10/4/62. Pres. on 2/28/63 and all rolls thru 2/28/65. Paroled at Appomattox C.H., VA on 4/9/65.

NEWBILL, JOHN: Pvt. Enl. on 3/12/62 as a substitute for Dr. John M. Evans. Absent sick in hosp. since 4/28/62 per 4/30/62 roll. In Petersburg Gen. Hosp. 10/22-28/62 and admitted to Chimborazo Hosp. #1 on 12/12/62, both hospitalizations for remittent fever. Transferred to Jackson Flying Art. on 10/4/62. Pres. on 2/28/63 and 4/30/63. Detailed to ordnance train on 6/20/63 per 8/31/63 and all subsequent rolls. In Chimborazo Hosp. #2 9/29-10/6/64 with rheumatism. Furloughed for 30 days on 10/6/64. Paroled at Appomattox. Ch., VA on 4/9/65. Res. of Middlesex Co.

O'BRIEN, ROBERT: 1st Lt. Enl. on 5/24/61 at Urbanna, VA in Co. B, 55th VA Inf. Under arrest on 12/22/61-1/62. AWOL on 2/25/62. Resigned on 3/3/62.

PARK, JOHN: Pvt. Enl. on 5/24/61 in Urbanna. Absent sick in Middlesex Co. since 2/28/62 per 4/30/62 roll.

PARKER, ALFRED H.: Pvt. Enl. on 5/24/61 at Urbanna, VA in Co. B, 55th VA Inf. Pres. on 4/30/62. Transferred to Jackson Flying Art. on 10/4/62. Pres. on 2/28/63, 4/30/63, 8/31/63, 10/31/63, 12/31/63, and 2/29/64 roll. WIA. In Gen. Hosp. #9 on 10/9-11/64, and in Chimborazo Hosp. #1 10/10-13/64 with wounds. Pres. on 10/31/64, 12/31/64, and 2/28/65 rolls.

PARKER, JOHN: Pvt. Enl. on 5/24/61 at Urbanna, VA in Co. B, 55th VA Inf. Detailed as a carpenter 11/61-1/62. Sick at home 3/1/62.

PARRON, RICHARD H.: Pvt. Enl. on 5/24/61 at Urbanna, VA in Co. B, 55th VA Inf. Pres. on 4/30/62. In Chimborazo Hosp. #3 8/17-9/12/62 for debilities.

PERRY, ELIJAH R.: Pvt. Enl. on 5/24/61 at Urbanna, VA in Co. B, 55th VA Inf. Sick at home 12/24/61-2/62. Pres. on 4/30/62.

PERRY, HARVEY D.: 7th Cpl. Enl. on 5/24/61 at Urbana, VA. Transferred from Middlesex Art. on 10/4/62. Pres. on 2/28/63 and 4/30/63 rolls as Pvt. Promoted to 7th Cpl. on 8/10/63. Pres. on 8/31/63, 10/31/63, and 12/31/63 rolls. Absent on furlough on 2/29/64. Transferred to C.S. Navy on 4/6/64 by S.O. 95, para 15, HQ, ANV.

PERRY, LEVI LUCAS: Postwar reference. D. 1908.

PERRY, R. B.: Postwar reference.

PIERCE, SAMUEL B.: Pvt. Enl. on 4/24/62 in Spotsylvania Co., $50 enlistment bounty due. Pres. on 4/30/62. Transferred to Jackson Flying Art. on 10/4/62, never reported per 2/28/63 roll.

POOR, JOHN A.: Pvt. Enl. in Co. D, 3rd VA Art. LDT. Transferred to Capt. Fleet's Battery (Middlesex Art.) on 5/14/62 by S.O. 21, HQ, 1st Brigade.

PRICE, SAMUEL B.: Pvt. Enl. on 4/24/62 in Spotsylvania Co. Transferred from Middlesex Art. Never reported.

PROFFITT, WILLIAM H.: 5th Cpl. Enl. on 5/1/62 in Richmond by Capt. Fleet or Capt. Johnson. Transferred to Jackson Flying Art. on 10/4/62. Pres. on 2/28/63 and all subsequent rolls thru 12/31/63. Absent, detail with wagons under Lt. Hunton on 2/29/64 as Pvt. Pres. on 10/31/64, 12/31/64, and 2/28/65 rolls as 5th Cpl.

RANSOM, HENRY T.: Pvt. Enl. on 2/22/62 in Richmond in Co. I, 3rd VA Art. LDT. (5/15/62 in Fredericksburg in Middlesex Art.) Transferred to Capt. Fleet's Battery (Middlesex Art.) on 5/14/62 by S.O. 21, HQ, 1st Brigade. Absent sick in hosp. on 2/28/63 and subsequent rolls until discharged by medical board on 9/5/63 for chronic rheumatism, described as 6'2", dark comp., brown eyes, black hair, occ. miner. Born in Cumberland Co. Res. of Garey Store, Buckingham Co., VA.

REVERE, ELIAS HENRY: Cpl. Enl. on 5/24/61 at Urbanna, VA in Co. B, 55th VA Inf. Appointed Cpl. on 4/28/62. Pres. 4/30/62 roll. Promoted to Cpl. on 4/28/62. Reassigned to the Ashland Art. on 10/4/62 by S.O. 209, Headquarters, Army of Northern Virginia. Pres. most of the war until KIA at Spotsylvania C.H., VA on 5/12/64, buried at old Spotsylvania C.H., VA in a box with a head board in a graveyard.

RILEE, GIDEON R.: Artificer. Enl. on 5/24/61 at Urbanna, VA in Co. B, 55th VA Inf. Pres. sick on 4/30/62.

ROACH, FRANK: Pvt. Enl. on 5/24/61 at Urbanna, VA in Co. B, 55th VA Inf. Sick at home winter 61- 62. Pres. on 4/30/62, $50 reenlistment bounty due.

ROACH, FRANK: Pvt. Enl. on 5/24/61 in Urbanna, VA. Transferred from Middlesex Art. on 10/4/62. Pres. on 2/28/63 and subsequent rolls thru 12/31/64. Absent on furlough on 2/28/65. In Chimborazo Hosp. #4 10/10-19/63 with intermittent fever.

ROANE, ADDISON M.: Sgt. Enl. on 5/24/61 at Urbanna, VA in Co. B, 55th VA Inf. Promoted to Sgt. on 4/28/62. Pres. on 4/30/62, described as age 23, blue eyes, light hair, light comp., 6', occ. clerk. Born in Gloucester Co., VA. Later served as enrolling officer (2nd Lt.) in 1st Congressional District.

ROANE, TIMOTHY: 1st Lt. Enl. on 5/24/61 at Urbanna, VA in Co. B, 55th VA Inf. Promoted to Jr. 1st Lt. on 4/28/62. Pres. on 4/30/62. S.O. 1383, dated 11/15/62, ordered him to report to Gen. French at Petersburg for duty in camp of paroled prisoners. Assigned as enrolling officer for King and Queen Co. on 2/29/64. Later detailed as enrolling officer, Alleghany Co., VA. On duty thru at least 11/30/64.

SAUNDERS, JOHN R.: Bugler. Enl. on 5/24/61 at Urbanna, VA. Transferred from Middlesex Art. on 10/4/62. Pres. on 2/28/63 and all subsequent rolls. Promoted to Bugler on 8/30/63 roll. Paroled at Appomattox C.H., VA on 4/9/65.

SAUNDERS, JOHN R.: Fifer. Enl. on 5/24/61 at Urbanna, VA in Co. B, 55th VA Inf. Elected fifer on 7/19/.61. Sick at home 10/27/61-12/61. Ceased to act as fifer on 4/1/62. Pres. on 4/30/62. Res. of Middlesex Co.

SAUNDERS, THOMAS J. J.: Pvt. Enl. on 4/3/62 at Fort Lowry, $50 enlistment bounty due. Pres. on 4/30/62.

SAUNDERS, THOMAS J.: Pvt. Enl. on 4/3/62 at Urbanna, VA (Fort Lowery, VA). Detailed as a nurse at Gordonsville Gen. Hosp. on 8/12/62. Transferred from Middlesex Art. on 10/4/62. Absent detailed to Hosp. at Gordonsville, VA on all rolls thru 2/29/64. In Gen. Hosp. #9 11/29- 30/63. Pres. on 10/31/64, 12/31/64, and 2/28/65 rolls. Paroled at Appomattox C.H., VA on 4/9/65.

SAUNDERS, WILLIAM J.: Cpl. Enl. on 5/24/61 at Urbanna, VA in Co. B, 55th VA Inf. Promoted to Cpl. on 4/28/62. Pres. on 4/30/62 roll. Promoted to Cpl. on 4/28/62. Reassigned to the Ashland Art. on 10/4/62 by S.O. 209, Headquarters, Army of Northern Virginia. Pres. on all rolls thru 2/28/65. Paroled at Appomattox C.H., VA on 4/9/65. D. 4/6/1888 in Pittsylvania Co., VA, age 56.

SEWARD, EDMUND W.: Sgt. Enl. on 5/24/61 at Urbanna, VA in Co. B, 55th VA Inf. Promoted to Sgt. on 4/28/62. Pres. on 4/30/62.

SEWARD, EDWARD W.: Jr. 2nd Lt. Enl. on 5/24/61 in Urbanna, VA. Transferred from Middlesex Art. on 10/4/62. Pres. on 2/28/63 and 4/30/63 as Pvt. Pres. on 8/31/63, 10/31/63, 12/31/63, and 2/29/64 rolls as Jr. 2nd Lt. NFR.

SEWARD, JOHN B.: Pvt. Enl. on 5/24/61 at Urbanna in Co. B, 55th VA Inf. Pres. sick on 4/30/62 roll. Reassigned to the Ashland Art. on 10/4/62 by S.O. 209, Headquarters, Army of Northern Virginia. Pres. on all rolls thru 2/28/65. Paroled at Appomattox C.H., VA on 4/9/65. Born in Essex Co., VA. Res. of Middlesex Co. Entered Old Soldier's Home, Richmond on 3/13/1901, age 58. D. 5/30/1908, buried in the Hollywood Cem., Richmond, VA.

SIBLEY, W. J.: Pvt. Enl. on 5/24/61 at Urbanna, VA. Transferred from Middlesex Art. on 10/4/62. "Deserted as soon as paid off" per 2/28/63 roll.

SIBLEY, WILLIAM J.: Pvt. Enl. on 5/24/61 at Urbanna, VA in Co. B, 55th VA Inf. Employed as regimental QM 1/1/62-1/31/62. Pres. sick on 4/30/62. In Chimborazo Hosp. #1 5/28- 6/17/62 with rheumatism.

SMITH, JAMES M.: Pvt. Enl. on 5/27/61 at Urbanna, VA in Co. B, 55th VA Inf. Pres. on 4/30/62. Transferred to Jackson Flying Art. on 10/4/62. Died per 2/28/63 roll. Died in Emory and Henry Hospital, Washington Co., VA in 1/63 per 8/31/63 roll.

SMITH, JOSEPH: Pvt. Enl. in Co. I, 3rd VA Art. LDT. Transferred to Capt. Fleet's Battery (Middlesex Art.) on 5/14/62 by S.O. 21, HQ, 1st Brigade.

SMITH, W. H.: Pvt. Enl. on 5/24/61 at Urbanna, VA. Transferred to Jackson Flying Art. on 10/4/62. Never reported, had actually died at Gordonsville, VA in 9/62.

SMITH, WILLIAM H.: Pvt. Enl. in Co. E, 3rd VA Art. LDT. Transferred to Capt. Fleet's Battery (Middlesex Art.) on 5/14/62 by S.O. 21, HQ, 1st Brigade.

SOUTH, ANDREW J.: Cpl. Enl. on 5/21/61 at Urbanna, VA in Co. B, 55th VA Inf. Detailed as carpenter constructing stables and forage house 12/61-1/62. Pres. on 4/30/62 roll. Appointed Cpl. on 4/28/62. Reassigned to the Ashland Art. on 10/4/62 by S.O. 209, Headquarters, Army of Northern Virginia. WIA at Gettysburg, PA on 7/2/63 in finger and knee. RTD by 4/30/64. Absent thru 2/28/65. Various hospitalizations noted in file.

SOUTH, ANDREW S.: Pvt. Enl. on 8/20/61 at Urbanna, VA in Co. B, 55th VA Inf. Pres. sick on 4/30/62. Transferred to Jackson Flying Art. on 10/4/62. Never reported per 2/28/63 roll. Absent sick since 11/62 per 4/30/63 roll.

STREET, ROBERT B.: 1st Sgt. Enl. on 6/15/61 at Urbanna, VA in Co. B, 55th VA Inf. Promoted to Cop. on 4/28/62. Pres. sick on 4/30/62. Transferred to Jackson Flying Art. on 10/4/62. Pres. on 2/28/63 and 4/30/63 rolls as Pvt. Promoted to Cpl. on 6/30/63. Promoted to 5th Sgt. on 8/10/63. Pres. on 8/31/63, 10/31/63, 12/31/63 and 2/29/64 rolls as 5th Sgt. Pres. on 10/31/64 and 12/31/64 as 1st Sgt. Absent sick in hosp. on 2/28/65. Paroled at Appomattox C.H., VA on 4/9/65.

SUTTON, JOHN W.: Sr. 2nd Lt. Enl. on ?. Dropped at reorganization on 4/28/62.

THOMAS, JOHN: Pvt. Enl. on ?. Co. H, 3rd VA Art. LDT. Transferred to Capt. Fleet's Battery (Middlesex Art.) on 5/14/62 by S.O. 21, HQ, 1st Brigade. Discharged from that unit on 9/20/62 at Winchester, VA. Died in service in Richmond, VA, described as age 42, 5'6", dark comp., dark eyes, black hair, occ. farmer. Ann E. Thomas, widow, filed claim for pay and allowances on 9/24/62. Born in Fluvanna Co.

THOMAS, SAMUEL P.: Pvt. B. 12/16/39. Enl. in Co. H, 3rd VA Art. LDT. Transferred to Capt. Fleet's Battery (Middlesex Art.) on 5/14/62 by S.O. 21, HQ, 1st Brigade. D. 12/27/20. Buried Mt. Vernon, UMC near Pt. Eastern, Va.

THORNTON, JAMES: Pvt. Enl. on 5/24/61 at Urbanna, VA in Co. B, 55th VA Inf. Pres. on 4/30/62. Transferred to Jackson Flying Art. on 10/4/62. Pres. on 2/28/63 thru 2/29/64 rolls. Absent detailed as a driver in QM Dept. on 10/31/64. Pres. on 12/31/64. Absent detailed as a driver in QM Dept. on 2/28/65. Paroled at Appomattox C.H., VA on 4/9/65.

THORNTON, WILLIAM T.: Pvt. Enl. on 5/24/61 at Urbanna, VA in Co. B, 55th VA Inf. Detailed to work on gun boats 2/27-3/31/62. Pres. on 4/30/62. Transferred to Jackson Flying Art. on 10/4/62. Pres. on 2/28/63 and all subsequent rolls. Paroled at Appomattox C.H., VA on 4/9/65.

THRIFT, WILLIAM T.: Pvt. Enl. on 7/20/61 at Urbanna, VA in Co. B, 55th VA Inf. Pres. sick on 4/30/62. Transferred to Jackson Flying Art. on 10/4/62. Pres. on all rolls from 3/28/63 thru 8/31/63. Absent in Hosp. in Culpeper Co. for chronic rheumatism since 10/11/63 per 10/31/63 roll. On sick furlough on 12/31/63, and 2/29/64 rolls. Pres. on 10/31/64, 12/31/64, and 2/28/65 rolls. Paroled at Appomattox C.H., VA on 4/9/65.

THROGG, WILLIAM: Pvt. Enl. in Co. H, 3rd VA Art. LDT. Transferred to Capt. Fleet's Battery (Middlesex Art.) on 5/14/62 by S.O. 21, HQ, 1st Brigade.

TRICE, ROBERT M. (N.): Pvt. Enl. on 5/24/61 at Urbanna, VA in Co. B, 55th VA Inf. Pres. on 4/30/62 roll. Reassigned to the Ashland Art. on 10/4/62 by S.O. 209, Headquarters, Army of Northern Virginia. Pres. on all rolls thru 2/28/65. Paroled at Appomattox C.H., VA on 4/9/65. Res. of Middlesex Co.

TRUMAN, WILLIAM P.: Pvt. Enl. in Co. H, 3rd VA Art. LDT. Transferred to Capt. Fleet's Battery (Middlesex Art.) on 5/14/62 by S.O. 21, HQ, 1st Brigade.

TURNER, JAMES W.: Pvt. Enl. in Co. H, 3rd VA Art. LDT. Transferred to Capt. Fleet's Battery (Middlesex Art.) on 5/14/62 by S.O. 21, HQ, 1st Brigade.

TUTON, WILLIAM: Pvt. Enl. on 5/27/62 at Urbanna, VA. Pres. on 4/30/62.

WALKER, ROBINSON D.: Pvt. Enl. on 5/24/61 at Urbanna in Co. B, 55th VA Inf. Pres. on 4/30/62 roll. Reassigned to the Ashland Art. on 10/4/62 by S.O. 209, Headquarters, Army of Northern Virginia. Pres. on all rolls thru 2/28/65. Paroled at Appomattox C.H., VA on 4/9/65. Res. of Middlesex Co.

WARE, WILLIAM S.: Pvt. Enl. on 5/24/61 at Urbanna, VA. Paid on 3/6/62. NFR.

WARREN, WILLIAM D.: Pvt. Enl. in Co. G, 3rd VA Art. LDT. Transferred to Capt. Fleet's Battery (Middlesex Art.) on 5/14/62 by S.O. 21, HQ, 1st Brigade.

WATT, ELISHA T.: Pvt. Enl. in Co. G, 3rd VA Art. LDT. Transferred to Capt. Fleet's Battery (Middlesex Art.) on 5/14/62 by S.O. 21, HQ, 1st Brigade.

WHITLOCK, J. M.: Pvt. Enl. in Co. H, 3rd VA Art. LDT. Transferred to Capt. Fleet's Battery (Middlesex Art.) on 5/14/62 by S.O. 21, HQ, 1st Brigade.

WOOD, JOHN M.: Pvt. Enl. on 3/17/62 at Fort Lowry in Co. B, 55th VA Inf. Pres. on 4/30/62. Later served in the Amherst Art., served until taken POW at Waynesboro, VA on 3/2/65, sent to Fort Delaware, where held until released on oath on 6/21/65. Res. of Middlesex Co., dark comp., dark hair, dark eyes, 5'10".

WOOD, NATHANIEL S.: Pvt. Enl. on 7/12/61 at Camp Sullivan in Co. B, 55th VA Inf. Pres. on 4/30/62. Later served in the Amherst Art. Taken POW at Waynesboro, VA on 3/2/65, sent to Fort Delaware, where held until released on oath on 6/21/65. Res. of Middlesex Co., dark comp., dark hair, dark eyes, 5'7".

WOOD, RICHARD T.: Pvt. Enl. on 12/1/61 at Urbanna, VA in Co. B, 55th VA Inf. Previously served in Co. B, 109th VA Militia. Pres. sick on 4/30/62. Most of service spent absent sick.

WOOLARD, WILLIAM JETER: Served in the Middlesex Art. per documents in the Amherst Art., in which he subsequently served.

ZILLS, JOHN: QM Sgt. Enl. on 5/24/61 at Urbanna, VA in Co. B, 55th VA Inf. Appointed QM Sgt. on 3/1/62 at Fort Lowry. Pres. on 4/30/62. Transferred to Jackson Flying Art. on 10/4/62. Absent sick on 2/28/63 and 4/30/63 rolls. AWOL since 1/20/63 per 8/31/63 roll. Pres. on 10/31/63, 12/31/63, 2/29/64, 10/31/64, 12/31/64, and 2/28/65 rolls.

ANDERSON, CLAUDIUS M.: Pvt. Son of John T. Anderson. Enl. 3/14/62, Hanover CH. Present through 4/30/62. Died of typhoid fever at Richmond, 7/12/62; cert. #12738, $53.40 due.

ANDERSON, EDMUND M.: Pvt. B. Hanover Co. Enrolled 5/21/61 at Beaver Dam Depot; age 20, student. Enl. 6/14/61, Jamestown. On furlough 2/7/62 - 3/13/62. Present 4/30/62. Transferred to Kirkpatrick's Co.; 5'8½", light comp., grey eyes, dark hair. KIA Cold Harbor.

ANDERSON, MARCELLUS M.: Pvt. B. Hanover Co. Enl. 3/11/62, Hanover. Discharged 8/11/62; age 43, 5'8", fair comp., grey eyes, brown hair, farmer. Disabled.

ASHWORTH, HENRY R.: Pvt. Enl. on 5/21/61 at Beaver Dam Depot, described as born in Lunenburg Co., carpenter, age 19, 5'10", light comp., brown hair, blue eyes. Reassigned to the Ashland Art. on 10/8/62 by S.O. 209, Headquarters, Army of Northern Virginia. Detailed as a teamster much of the war. Paroled at Burksville, VA on 4/23/65.

BASYE, WILLIAM: 2nd Lt. See Co. E, 21st Va. Militia. Enl. 10/19/61. Enl. 10/19/61, Gloucester Co. Resigned 5/1/62 due to election as 2nd Lt. of Nelson's Co. 4/25/62. Assigned as Lt. in Kirkpatrick's Co., 10/22/62. Wounded 7/20/64 and died as a result at Chimbo., 8/1/64.

BERKELEY, HENRY ROBINSON: Pvt. Enl. 5/21/61 at Beaver Dam Depot; age 21, student. Transferred to Kirkpatrick's Co., 10/62. D. 1/16/18, age 70, Orange, VA.

BERKELEY, JOHN HILL: Pvt. Enl. 5/21/61, Beaver Dam Depot. Transferred to Kirkpatrick's Co., 10/62. Took oath in Ashland, 4/23/65.

BERKELEY, JOHN L.: Pvt. Enl. 3/10/62, Hanover Co. Transferred to Kirkpatrick's Co. Paroled in Ashland, 4/25/65.

BLAKE, JACOB: Pvt. Enl. on 4/1/62 in Richmond, VA. Bounty recd. ($50), 6/13/62. Reassigned to the Ashland Art. on 10/8/62 by S.O. 209, Headquarters, Army of Northern Virginia. AWOL on 10/31/62 roll, deserted prior to reporting to Ashland Art.

BOWLES, R. N.: Postwar reference. Amherst p.l.

CHILDRESS, JOHN W.: Pvt. Enl. on 3/25/62 in Hanover Co. Deserted on 8/20/62.

COOKE, CHARLES: 1st Lt. Enl. 5/21/61, Beaver Dam Depot; age 29, lawyer. Took leave per S.O. 273/6, 12/23/61. NFR.

COOKE, JOHN McP.: Pvt. B. Hanover Co. Enl. 5/21/61, Beaver Dam Depot; age 19, student. Absent on sick leave 8/31/61. On furlough 2/28/62. Present 4/30/62. Transferred to Co. G, 4th Va. Cav. Per S.O. #172/4, 8/4/62. 5'9", florid comp., blue eyes, black hair.

CORKER, JOHN COALMAN: Pvt. Enl. on 5/21/61 in Hanover Co., described as born in Louisa Co., 5'7", age 24, light comp., light hair, blue eyes, farmer. Furloughed 2/28/62. Present 4/30/62. Reassigned to the Ashland Art. on 10/8/62 by S.O. 209, Headquarters, Army of Northern Virginia. AWOL on 10/31/62 roll. Arrested in Dept. of Henrico 11/15/62. Pres. on all rolls until shown as being on furlough on 2/28/65 roll. WIA at some point. Paroled at Ashland, VA on 4/21/65. Age 37, living in Hanover Co. D. 12/10/1905 of coronary thrombosis, buried in Elon Baptist Church Cem., Hanover Co. Widow applied for pension in Hanover Co. in 1909.

CORKER, JOSEPH F.: Pvt. Enl. on 5/21/61 at Beaver Dam Depot, described as born in Louisa Co., age 30, 5'6", light comp., light hair, blue eyes, farmer. Furloughed 2/28/62. Present 4/30/62. Reassigned to the Ashland Art. on 10/8/62 by S.O. 209, Headquarters, Army of Northern Virginia. Pres. or accounted for rest of war. D. 4/25/1888 near Beaver Dam or neuralgia of the heart. Buried in the Elon Baptist Church Cem., Hanover Co.

DEALES, WILLIAM P.: Pvt. Enl. on 3/25/62 in Hanover Co. Deserted from Yorktown on 4/10/62. Reassigned (but not present) to the Ashland Art. on 10/8/62 by S.O. 209, Headquarters, Army of Northern Virginia. Detailed in Staunton Hosp fall 1862. Taken POW at Mossy Creek, TN, sent to Camp Chase, OH and then to Rock Island, IL. NFR.

DUKE, LUTHER W.: Pvt. Enl. on 5/21/61 at Beaver Dam Depot, described as age 20, born in Louisa Co., 5'10-", light comp., light hair, brown eyes, occ. mechanic. Furloughed 2/28/62. Pres. 4/30/62. Reassigned to the Ashland Art. on 10/8/62 by S.O. 209, Headquarters, Army of Northern Virginia. Appointed QM Sgt. on 4/23/63. Pres. thru 2/28/65 roll. Paroled at Appomattox C.H., VA on 4/9/65.

DUKE, THOMAS A.: Pvt. Enl. on 7/23/61 at Jamestown, James City Co., VA, described as age 30, teacher, born in Louisa Co., light comp., dark hair, hazel eyes. Furloughed 2/28/62. Detailed as clerk on 4/30/62 roll. Appointed QM Sgt. Reassigned to the Ashland Art. on 10/8/62 by S.O. 209, Headquarters, Army of Northern Virginia. Detailed to QM Dept. most of the war. Paroled at Appomattox C.H., VA on 4/9/65. Res. of Hanover Co.

DURVIN, JOHN W.: Pvt. Enl. 5/21/61, Beaver Dam Depot; age 41, mechanic. Transferred To Kirkpatrick's Co.

ESKEW, SILAS: Pvt. Pres. 3/15/62. Furloughed from soldier's home 7/11/62. Transferred To Woolfolk's Co.

ESKEW, SYLVANUS: Pvt. Enl. on 3/15/62 in Hanover Co. Deserted at Yorktown 4/10/62. AWOL in 10/62. Arrested by Co. C, Bogg's Bttn., 12/2/62 and charged $30. Transferred to Graham's Battery (Petersburg Horse Art.) on 3/1/63 by Gen. Samuel Cooper. No record w/Graham's Battery. See also Woolfolk's Co.

FARNER, JOHN: Postwar reference. Southern Historical Papers.

FLEMING, ROBERT E.: Pvt. Enl. 3/25/62, Hanover. Absent sick 6/30/62. At Chimbo. #2 w/typhoid, 8/20/62. Furloughed 8/28, for 35 days. Paroled 4/22/65, Richmond.

FLIPPO, WILLIAM S.: Cpl. B. 9/24/32. Enl. on 5/21/61 in Richmond, VA, described as born in Caroline Co., farmer, age 28, 5'7-", light comp., dark hair, grey eyes, farmer. Reassigned to the Ashland Art. on 10/8/62 by S.O. 209, Headquarters, Army of Northern Virginia. Pres. thru 2/28/65 roll. Paroled at Appomattox C.H., VA on 4/9/65. Res. of Hanover Co. Living in Caroline Co. in 1870, age 38. D. 10/11/1892, buried in the Bethany Church Cem., Caroline Co., VA.

FONTAINE, WILLIAM MORRIS: 1st Lt. B. Louisa Co. Enl. 5/21/61, Beaver Dam Depot; age 26, teacher. 1st Lt. 4/30/62. D. 4/29/13, Charlottesville, VA.

FUQUA, JOSEPH: Pvt. Appointed hospital seward, S.O.# 210, 8/28/62 to Surgeon Taliaferro at Farmville.

GENTRY, WILLIAM J.: Pvt. Enl. on 5/21/61 in Hanover Co., described as born in Caroline Co., occ. mechanic, age 28, 5'11", dark comp., black hair, grey eyes. Reassigned to the Ashland Art. on 10/8/62 by S.O. 209, Headquarters, Army of Northern Virginia. WIA at Fredericksburg on 12/13/62. Transferred to Co. B, 30th VA Inf. by S.O. 283/31 in exchange for J. E. Goldsby. Applied for pension in Caroline Co. in 1902.

GILSON, JAMES D.: Pvt. B. Cavan Co., Ireland. Enl. 5/21/61, Beaver Dam Depot; age 34, mechanic. Discharged as not subject to the conscription act, 7/17/62; age 39, 6', dark comp. and hair, grey eyes, blacksmith.

GRAVES, BERNARD B.: 1st Sgt. Enl. 5/21/61, Beaver Dam Depot; age 24, farmer. Elected Cpl. Furloughed 8/31/61. 1st Sgt. (6/30/62).

GRIFFIN, JAMES A.: Postwar reference. B. 1845, Nelson County, Va. Came to Beaumont, Jefferson Co., Texas in 1892. D. 2/6/30, Austin, Texas. Buried in State Cem., Section F, Row U.

GRIMES, GEORGE WASHINGTON: Cpl. B. Spotsylvania Co. Enl. 5/21/61, Beaver Dam Depot; age 23, farmer. Furloughed 2/28/62. Pres. 4/30/62. Cpl. 6/15/62. 5'6", light comp. and hair, grey eyes, also listed as a sailor.

GRIMES, JOSEPH D.: Pvt. Enl. on 5/21/61 in Hanover Co., described as born in Hanover Co., farmer, age 25, 5'6", dark comp., black hair, black eyes. Furloughed 2/28/62. Pres. 4/30/62. Reassigned to the Ashland Art. on 10/8/62 by S.O. 209, Headquarters, Army of Northern Virginia. AWOL on 10/31/62. Arrested on 11/15/62 in Dept. of Henrico. Pres. 12/31/62 thru 2/28/65 rolls. Paroled in Ashland, VA on 4/21/65. Res. of Hanover Co.

GRIMES, WILLIAM T.: Pvt. Enl. 3/10/62, Hanover CH. Transferred to Kirkpatrick's Co.

HAINES, GEORGE WASHINGTON: Pvt. Enl. on 3/10/62 in Hanover Co. Reassigned to the Ashland Art. on 10/8/62 by S.O. 209, Headquarters, Army of Northern Virginia. Absent sick or detailed much of the time until taken POW in TN. on 1/24/64. No record of incarceration. In Jackson Hosp., Richmond, VA on 3/6/65 for debility, given 30 day furlough.

HAINES, JAMES H.: Pvt. Enl. on 3/10/62 in Hanover Co. Reassigned to the Ashland Art. on 10/8/62 by S.O. 209, Headquarters, Army of Northern Virginia. Pres. on all subsequent rolls. Paroled at Appomattox C.H., VA on 4/9/65. Res. of Hanover Co. Bro. of John H.

HAINES, JOHN H.: Pvt. Enl. on 3/10/62 in Hanover Co. Reassigned to the Ashland Art. on 10/8/62 by S.O. 209, Headquarters, Army of Northern Virginia. Pres. on all subsequent rolls. Paroled at Appomattox C.H., VA on 4/9/65. Res. of Hanover Co.

HAINES, JOHN W.: Pvt. Enl. 3/10/62. At Petersburg G.H. 6/12 - 7/7/62. NFR.

HALL, CHARLES F.: Pvt. Enl. 5/21/61, Beaver Dam Depot; age 19, farmer. Furloughed 2/28/62. Pres. 4/30/62. Discharged 7/11/62; age 20, 5'10", light comp., dark hair and eyes.

HALL, CHARLES M.: Pvt. B. Hanover. Son of Lewis E. Hall. Enl. 5/21/61, Beaver Dam Depot; age 36, farmer, 6' 2½", dark comp., hazel eyes. Furloughed 2/28/62. Pres. 4/30/62. Arrested L.R.#319, 11/15/62. D. 6/2/62; cert. #22378, $85.50 due.

HALL, FRANKLIN S.: 1st Cpl. Enl. 7/1/61. 1st Cpl. (2/28/62). Transferred To Pollick's Fredericksburg Artillery per S.O. #104/27, 5/6/62.

HALL, GEORGE WASHINGTON: Pvt. B. Louisa Co. Enl. 5/21/61, Beaver Dam Depot; age 33, farmer, 5'11", light comp., blue eyes, light hair. Furloughed 2/28/62. Pres. 4/30/62. Transferred To Kirkpatrick's Co.

HALL, JULIAN H.: Pvt. B. Hanover Co. Enl. 5/21/61, Beaver Dam Depot; age 22, farmer, 5'10½", dark comp. and hair, blue eyes. Furloughed 2/28/62. Pres. 4/30/62. Transferred To Kirkpatrick's Co.

HALL, MACH: Pvt. Enl. 5/21/61, Beaver Dam Depot; age 31, farmer, 6'2", dark comp. and hair, grey eyes. Furl. 2/28/62. Pres. 4/30/62. Transferred To Kirkpatrick's Co.

HALL, THOMAS H.: Pvt. Enl. ? Paroled in Richmond, 4/25/65.

HALL, THOMAS S.: Pvt. B. Hanover Co. Enl. 5/21/61, Beaver Dam Depot; age 24, gentleman, 6', light comp., dark hair, grey eyes. Furl. 2/28/62. Pres. 4/30/62. Transferred To Kirkpatrick's Co.

HALL, WILLIAM W.: Pvt. Enl. 5/21/61, Beaver Dam Depot; age 42, mechanic/carpenter, 5'11", light comp., blue eyes, light hair. Furloughed 2//28/62. Pres. 4/30/62. Transferred To Kirkpatrick's Co.

HARPER, BENJAMIN J.: Pvt. Enl. 5/21/62, Beaver Dam Depot; age 25, farmer, 5'10", light comp., grey eyes, dark hair. Furloughed 2/28/62. Pres. 4/62. At Chimbo. #5 with fever, 4/25/62. D. 5/3/62. Claim filed by mother, S. Harper, 9/1/63. Cert. #22545, $49.20 found due 2/4/65.

HARRIS, ANDREW JACKSON: Postwar reference. Amherst p.l.

HARRIS, BENJAMIN FRANKLIN: Postwar reference. KIA Sharpsburg.

HARRIS, ELISHA W.: Pvt. Enl. 3/15/62. Transferred To Kirkpatrick's Co.

HARRIS, SAMUEL: Postwar reference. KIA Sharpsburg.

HOLLOWAY, ALEXANDER J.: Sgt. Enl. 5/21/61, Beaver Dam Depot; age 20, farmer, 5'9 1/4", grey/hazel eyes, light/brown hair, dark comp. Furloughed 2/28/62. Pres. 4/62. Absent sick w/fever in Chimbo. #5, 6/10 - 6/16/62. Transferred To Kirkpatrick's Co.

HUFF, ANDREW JACKSON: Postwar reference. Amherst p.l.

JACKSON, THOMAS E.: Pvt. Enl. on 6/12/61 or 7/12/61 at Jamestown, James City Co., VA, described as born in Louisa Co., farmer, age 18, light comp., dark hair, black eyes. Furloughed 2/28/62. Pres. 4/62. Reassigned to the Ashland Art. on 10/8/62 by S.O. 209, Headquarters, Army of Northern Virginia. Sick on 10/31/62 roll. Pres. on subsequent rolls thru 2/28/65. Paroled at Appomattox C.H., VA on 4/9/65. Res. of Hanover Co.

JOHNSON, ANDREW J.: Pvt. Enl. on 3/14/62 in Hanover Co. Reassigned to the Ashland Art. on 10/8/62 by S.O. 209, Headquarters, Army of Northern Virginia. Pres. on all rolls thru 2/28/65. Paroled at Appomattox C.H., VA on 4/9/65. Res. of Hanover Co.

JONES, GEORGE W.: Pvt. Enl. on 3/10/62 in Hanover Co. Reassigned to the Ashland Art. on 10/8/62 by S.O. 209, Headquarters, Army of Northern Virginia. Absent sick 10/31/62. Deserted 9/17/63 at Petersburg. Court-martialed 9/25/63. Taken POW ion 5/25/64 at Hanover Junction, gave himself up to General. Wilson at North Anna River, claimed to have been conscripted in 8/1863. Sent to Point Lookout, MD and then to Elmira, NY where held until he died on 9/7/64 of chronic diarrhea. Buried in grave #196, Woodlawn National Cem.

LOWRY, JAMES: Pvt. Enl. on 3/10/62 in Hanover Co. Reassigned to the Ashland Art. on 10/8/62 by S.O. 209, Headquarters, Army of Northern Virginia, but never reported, deserted. Paroled 5/8/65, Beaver Dam Depot.

LOWRY, JOHN R.: Pvt. Enl. on 3/25/62 in Hanover Co. Reassigned to the Ashland Art. on 10/8/62 by S.O. 209, Headquarters, Army of Northern Virginia. AWOL on transfer. Pres. on 2/28/63 thru 2/28/65 rolls. Living in Hanover Co. in 1870, then age 48. Applied for pension in 1900, then age 72. D. 11/23/1910 near Beaver Dam Depot, Hanover Co.

LOWRY, LUTHER N.: Pvt. Enl. 5/21/61, Beaver Dam Depot; age 18, farmer. Transferred To Kirkpatrick's Co.

LOWRY, RICHARD EDWARD: Pvt. Enl. on 7/22/61 at Jamestown, James City Co., VA. Pres. on 8/31/61 roll. Status not stated on 4/30/62 roll. Reassigned to the Ashland Art. on 10/8/62 by S.O. 209, Headquarters, Army of Northern Virginia. Absent sick on 3/1/63. Admitted to Chimborazo Hosp. #3 9/28/64. Honorably retired to the Invalid Corps on 12/12/64 from chronic rheumatism. Applied for pension in Hanover Co. in 1900. Died on 4/14/1906 in Hanover Co.

LUCK, FREDERICK W.: Pvt. Enl. on 3/25/62 in Hanover Co. Reassigned to the Ashland Art. on 10/8/62 by S.O. 209, Headquarters, Army of Northern Virginia. Deserted at Millwood on 7/25/63. Paroled at Ashland, VA on 4/25/65. Res. of Hanover Co. Living in Louisa Co. in 1870, age 32. Died on 4/28/1919, buried in the Elon Baptist Church Cem., Hanover Co.

LUCK, SAMUEL A.: Pvt. Enl. 5/21/61, Beaver Dam Depot; age 28, farmer. D. in Yorktown, 2/11/62.

LUCK, WILLIAM J.: Pvt. B. Louisa Co. Enl. 5/21/61, Beaver Dam Depot; age 24, farmer, 5'7½", light comp., blue eyes, dark hair. Furloughed (2/28/62). Pres. (4/62). Died in service, date unknown. Widow, Credalia F. Luck, filed claim 6/22/63.

MADDOX, FRANKLIN L.: Pvt. B. Hanover Co. Enl. 5/21/61, Beaver Dam Depot; age 18, farmer/student, 5'10½", light comp. and hair, grey eyes. Absent on leave (6/30/61). Pres. (8/31/61). Absent sick 12/1/61. Discharged 9/20/62. Drew Tennessee (Franklin Co.) pension (#S13605).

MALLORY, COLUMBUS J.: Pvt. Enl. on 3/25/62 in Hanover Co. Reassigned to the Ashland Art. on 10/8/62 by S.O. 209, Headquarters, Army of Northern Virginia. Pres. on all subsequent rolls thru 2/28/65. Paroled at Appomattox C.H., VA on 4/9/65. Res. of Hanover Co. Applied for pension in Hanover Co. in 10/1904, then age 65.

MALLORY, HALLICK R.: Pvt. B. Hanover Co. Enl. 7/30/61, Jamestown. Absent sick 10/1/61. Discharged with chronic rheumatism, 7/62; age 19, dark hair, eyes and comp., farmer.

MALLORY, HORATIO G.: Pvt. Enl. on 7/30/61 at Jamestown, James City Co., VA. Furloughed (2/28/62). Pres. (4/62). Reassigned to the Ashland Art. on 10/8/62 by S.O. 209, Headquarters, Army of Northern Virginia. Pres. on all rolls thru 12/31/64. Detailed as shoemaker 2/28/65 roll. Applied for pension in Hanover Co. in 1904, age 64. Admitted to Old Soldier's Home, Richmond, VA on 5/31/1920, age 80. D. 8/31/1921.

MALLORY, FELIX G.: Pvt. Enl. 7/19/61, Jamestown. Discharged 11/4/61.

MALLORY, THOMAS B.: Pvt. Enl. on 3/25/62 in Hanover Co. Reassigned to the Ashland Art. on 10/8/62 by S.O. 209, Headquarters, Army of Northern Virginia. Pres. on all rolls thru 2/28/65. Paroled at Appomattox C.H., VA on 4/9/65. Res. of Hanover Co.

MASON, JOHN T.: Pvt. Enl. on 3/12/62 in Hanover Co. Reassigned to the Ashland Art. on 10/8/62 by S.O. 209, Headquarters, Army of Northern Virginia. Discharged on 3/1/63. Lynchburg p.l.

MASON, RICHARD O.: Pvt. Enl. on 3/25/62 in Hanover Co. Reassigned to the Ashland Art. on 10/8/62 by S.O. 209, Headquarters, Army of Northern Virginia. WIA at Gettysburg, PA on 7/2/63. Retired from service by Medical board due to gunshot wound to left knee, left thigh and right leg amputation 3" below the knee. Extensive hospitalization in Union and Confederate hospitals. Received benefits of Assn. for Relief of Maimed Soldiers. Received pension in 1888, res. of Hanover Junction, then age 60.

MILLS, SAMUEL: Pvt. Enl. on 5/21/61 in Hanover Co. Furloughed (2/28/62). Pres. (4/62). Reassigned to the Ashland Art. on 10/8/62 by S.O. 209, Headquarters, Army of Northern Virginia. WIA and taken POW at Gettysburg, PA, sent to Hammond U.S. Army Gen. Hosp. with flesh wound in left leg. Exchanged on 3/6/64. Hospitalized extensively after exchange. Taken POW on 3/19/65 on South Anna River, sent to Fort Monroe and then to Point Lookout, MD. Released on 6/15/65 on oath. Described as 5'9", fair comp., brown hair, blue eyes. Res. of Kilbys, Hanover Co., VA. Applied for pension on 10/16/1889, age 63.

MOSEBY, JOHN T. P.: Pvt. Enl. on 11/18/61 at Yorktown. Reassigned to the Ashland Art. on 10/8/62 by S.O. 209, Headquarters, Army of Northern Virginia. KIA at Chancellorsville, VA on 5/1/63.

MOSBY, PHILIP SAMUEL: Pvt. Enl. on 3/25/62 in Hanover Co. Reassigned to the Ashland Art. on 10/8/62 by S.O. 209, Headquarters, Army of Northern Virginia. Pres. on 10/31/62 roll. Sick on 2/28/63 roll. Detailed to Medical Battn. Discharged on 11/12/64 under writ of habeas corpus. Age 45, carpenter, living in Hanover Co., VA in 1870.

MURPHY, JAMES: Pvt. Enl. 5/21/61, Beaver Dam Depot; age 22, farmer/laborer, 5'10", light comp., dark hair and eyes. Furloughed (2/28/62). Pres. (4/62).

NELSON, GEORGE WASHINGTON: Captain. Enl. 5/21/61, Beaver Dam Depot; age 21, teacher, 5'11", florid comp., blue eyes, brown hair. Resident of Hanover Co. 1st Sgt., (6/30/61). 2nd Lt., (2/28/62). Captain commanding, 4/30/62. Major, 5/26/62. Captured at New Market, Va., 10/63. Sent to Wheeling, W. Va. (Athenium) 10/24/63. To Camp Chase, Ohio, 10/31/63. To Pt. Lookout, Md. 4/22/64. At Hammond, U.S.A. hospital, Pt. Lookout with chronic dysentery, 4/26/64. To Ft. Delaware, 6/23/64. D. 5/30/1903, Warrenton, VA.

NELSON, ROBERT C.: Pvt. Enl. 5/21/61, Beaver Dam Depot; age 19, student. NFR.

NELSON, WILLIAM: Captain. B. 12/14/1808, Yorktown, Va. Enl. 5/21/61; age 52, farmer. Absent on leave (6/30/61). Pres. (8/31/61). Dropped at reorganization on 4/20/62. Appointed Major of Artillery on 5/26/1862. Promoted to Lt. Col. on 3/3/63 and to Col. on 2/18/65. Sometimes described as commanding the 31st VA Light Artillery Battalion, actually known as Nelson's Battn. Lt. Col. 2/65. D. 4/17/1892, never married.

NUCKOLLS, ALPHEUS B.: Cpl. Enl. on 5/21/61 at Beaver Dam Depot; age 21, farmer. Reduced to ranks 8/5/61. Absent sick (8/31/61). On sick leave on 2/28/62 roll Reassigned to the Ashland Art. on 10/8/62 by S.O. 209, Headquarters, Army of Northern Virginia. Pres. on 10/31/62 roll. Sick on 6/3/63. Slightly WIA in back on 6/23/64. Pres. 12/31/63 thru 2/28/65 rolls. Paroled at Ashland, VA on 4/21/65. Res. of Hanover Co., VA.

O'BRIEN, PATRICK: Pvt. Enl. on 5/23/61, Beaver Dam Depot; age 24, mechanic. Described as being born in Ireland, age 27, occ. laborer, 5'8", dark comp., dark hair, hazel eyes. Reassigned to the Ashland Art. on 10/8/62 by S.O. 209, Headquarters, Army of Northern Virginia. Pres. thru 2/28/63. Deserted. Apprehended, RTD, Pres. thru 2/28/65 roll. Taken POW at Dinwiddie C.H., VA on 4/13/65, came in, gave himself up, asked to be paroled on terms of general surrender. Died on 3/28/1879 at Beaver Dam, Hanover Co. of heart disease.

O'BRIEN, PHILIP: Musician. Enl. 3/21/62, Richmond. Reassigned to the Ashland Art. on 10/8/62 by S.O. 209, Headquarters, Army of Northern Virginia. Pres. most of the rest of the war.

OLIVER, JOHN E.: Pvt. B. in Caroline Co., son of Reuben Oliver. Enl./ 7/22/61, Jamestown. Furloughed (2/28/62); age 20, 5'7 3/4", mason, light hair and comp., blue eyes. Died in service, date unknown. Claim filed 7/17/63, cert. #12303. $52.00 found due, 2/10/64.

PARISH, ELIHU: Pvt. Enl. 3/8/62, Richmond. Transferred To Kirkpatrick's Co.

PATE, MATTHEW W.: Pvt. Enl. 5/21/61, Beaver Dam Depot; age 35, farmer. Reassigned to the Ashland Art. on 10/8/62 by S.O. 209, Headquarters, Army of Northern Virginia. Deserted on 5/20/64 near Hanover Junction. Living in Hanover Co. in 1870 then age 46.

PERKINS, GEORGE WASHINGTON: Pvt. Enl. 5/21/61, Beaver Dam Depot; age 24, mechanic. Furloughed (2/28/62). Pres. (4/30/62).

PERKINS, JAMES H.: Pvt. Enl. on 5/21/61, Beaver Dam Depot. Described as farmer, age 25, 5'6", light comp., blue eyes, occ. mechanic. Reassigned to the Ashland Art. on 10/8/62 by S.O. 209, Headquarters, Army of Northern Virginia. Pres. on all subsequent muster rolls. Paroled at Ashland, VA on 4/21/65. Res. of Hanover Co.

PERKINS, JOSIAH: Pvt. Enl. 5/21/61, Beaver Dam Depot. Described as being born in Hanover Co., age 22, farmer, 5'7-", light comp., light hair, hazel eyes. Furloughed (2/28/62). Pres. (4/62).Reassigned to the Ashland Art. on 10/8/62 by S.O. 209, Headquarters, Army of Northern Virginia. WIA in arm and leg on 7/2/63 at Gettysburg, PA. Taken POW on 7/4/63 at Cashtown, PA, sent to Hammond Hosp., Chester, PA. Exchanged. Paroled at Appomattox C.H., VA on 4/9/65. Res. of Hanover Co. Applied for pension in 5/1894 then age 55, in Hanover Co.

PERKINS, ROBERT M.: Pvt. B. Hanover Co. Enl. 7/25/61, Jamestown. Furloughed (2/28/62); age 19, 6'1½", light comp., hazel eyes, dark hair, farmer. Pres. (4/62). Absent sick (6/30/62).

PERKINS, THOMAS F.: Pvt. B. Hanover Co. Enl. 5/21/61, Beaver Dam Depot; age 21, merchant. Furloughed (2/28/62); 5'9 3/4", light hair and comp., blue eyes. Pres. (4/62).

PERKINS, THOMAS N.: Pvt. B. Hanover Co. Enl. 5/21/61, Beaver Dam Depot; age 26, farmer. Furloughed (2/28/62); 5' 9½", light comp. and hair, grey eyes, farmer. Pres. (4/62).

PHILLIPS, ANDREW JACKSON: Pvt. Enl. 7/16/61, Jamestown. NFR.

POINTS, LEONIDAS: Sgt. Major. Enl. 5/21/61, Beaver Dam Depot; age 27, teacher. Sgt. Major, 8/27/61. Pvt. (2/28/62). At Chimbo. #5 w/diarrhea, 4/62. Reassigned to the Ashland Art. on 10/8/62 by S.O. 209, Headquarters, Army of Northern Virginia. Pres. until commissioned Lt. of Ordinance on 2/28/63. Acting 1st Lt. and Ord. Officer, Garnett's Battalion, 3/1/63. Captain, 10/6/63 to Col. J. Gorgas. Surrendered at Appomattox C.H. Living in 1894.

RAGLAND, CHARLES L.: Pvt. Enl. 5/21/61, Beaver Dam Depot; age 19, farmer. Described as born in Hanover Co., farmer, age 20, 6'-", light comp., black hair, gray eyes. Furloughed (2/28/62). Pres. (4/62). Reassigned to the Ashland Art. on 10/8/62 by S.O. 209, Headquarters, Army of Northern Virginia. Pres. on all rolls thru 2/28/65. Paroled at Appomattox C.H., VA on 4/9/65. Res. of Hanover Co.

RAGLAND, PETTUS W.: Pvt. Enl. 5/25/61, Beaver Dam Depot. Described as farmer, age 25. In Chimborazo Hosp.7/23/62 with debility. Reassigned to the Ashland Art. on 10/8/62 by S.O. 209, Headquarters, Army of Northern Virginia. Died in Richmond, VA Hosp. on 7/15/64.

RICKETTS, ____: Sgt. Postwar reference. Lynchburg pension list.

RODGERS, J. J.: Pvt. Enl. ? Paroled 4/22/65, Ashland.

ROUNQUIST, WILLIAM P.: Pvt. Enl. 7/25/61, Jamestown. Furloughed (2/28/62); age 19, light hair and comp., blue eyes. Pres. (4/30/62). Transferred to Kirkpatrick's Co.

ROUZIE, EDWARD J.: Sgt. Enl. 5/21/61, Beaver Dam Depot; age 28, medical doctor. Detailed to hospital service, 7/27/61. On rolls thru 4/30/62. Reassigned to the Ashland Art. on 10/8/62 by S.O. 209, Headquarters, Army of Northern Virginia. AWOL from transfer until 2/28/63. Deserted on 4/19/63 at Carmel Church, Caroline Co., VA. Living in Hanover Co., VA in 1870, age 36.

SAVAGE, THOMAS H.: Pvt. Enl. 12/3/61, Matthews C.H. Furloughed (2/28/62). Pres. (4/62). At Chimbo. #4 with fever 4/25 - 5/29/62.

SIMMS, SAMUEL H. C.: Pvt. B. Hanover Co. Enl. 5/21/61, Beaver Dam Station; age 19, farmer. Described as farmer, age 20, 5'9-", light comp., dark hair, grey eyes. Furloughed (2/28/62); 5'9½", light comp., grey eyes, dark hair. Pres. (4/62). Reassigned to the Ashland Art. on 10/8/62 by S.O. 209, Headquarters, Army of Northern Virginia. Pres. on all later muster rolls. Paroled at Appomattox C.H., VA on 4/9/65. Res. of Hanover Co.

SMITH, BENNETT L.: Pvt. Enl. on 7/10/61, Jamestown. Described as born in Hanover Co., mechanic, age 18, 5'9-", light comp., dark hair, blue eyes. Furloughed (2/28/62). Pres. (4/62). Reassigned to the Ashland Art. on 10/8/62 by S.O. 209, Headquarters, Army of N. Virginia. Slightly WIA at Fredericksburg, VA on 12/13/62. Pres. on all rolls thru 2/28/65. Court- Martialed Gen. Orders #90, 9/25/63. Paroled at Appomattox C.H., VA on 4/9/65. Res. of Hanover Co.

SNEAD, JAMES HENRY: Pvt. Enl. 8/5/61, Jamestown. Pres. on 8/31/61, 2/28/62 and 4/30/62 muster rolls. Reassigned to the Ashland Art. on 10/8/62 by S.O. 209, Headquarters, Army of Northern Virginia. WIA at Gettysburg, PA on 7/2/63. Taken POW at Cashtown, PA on 7/5/63, sent to Chester, PA hosp. Exchanged. In Chimborazo Hosp. #1 9/24/63. Extensively hospitalized remainder of the war. Paroled on 4/26/65. Res. of Albemarle Co. D. 4/9/1866 in Albemarle Co., VA. Widow filed pension application.

SOUTHARD, JAMES T.: Sgt. Enl. on 7/19/61, Jamestown. Described as born in Hanover Co., farmer, age 24, 5'10", light comp., dark hair, blue eyes. Furloughed (2/28/62). Pres. (4/62). Absent sick on 6/30/62 roll. Transferred to Capt. Thomas J. Kirkpatrick's Company Light Artillery. Taken POW at Gettysburg, PA on 7/3/63. Sent to Chester, PA Gen. Hosp. Exchanged at City Point, VA on 8/17/63. Reassigned to the Ashland Art. Pres. thru 2/28/65 roll. Paroled at Appomattox C.H., VA on 4/9/65. Res. of Hanover Co. Died 1893 of apoplexy, widow filed pension application in 1900.

SPICER, ROBERT ANDERSON: Pvt. Enl. on 5/12/61 in Hanover Co. Described as born in Missouri, occ. mechanic, age 18, 5'10-", light comp., light hair, blue eyes. Absent on leave (6/30/61). Furloughed (2/28/62). Pres. (4/62). Absent sick (6/30/62). Reassigned to the Ashland Art. on 10/8/62 by S.O. 209, Headquarters, Army of Northern Virginia. Transferred to Kirkpatrick's Battery on 10/13/62. AWOL on 10/31/62 roll. Discharged on surgeon's certificate on 10/13/62. Taken POW during Stoneman's Raid. Paroled at Thompson's X Roads on 6/2/63, due to lack of transportation.

STANLEY, LUCIUS C.: Pvt. B. Hanover Co. Enl. 5/21/61, Beaver Dam Depot; age 27, farmer. Furloughed (2/28/62); age 28, 6'2½", light comp. and hair, blue eyes. Pres. (4/62). At Chimbo. #5 with fever, 6/10/62. Sent to Camp Winder, 6/27/62. At Chimbo#5, 8/28/62. Furloughed for 40 days. Alive in 1885.

STAPLES, THOMAS JEFFERSON: Postwar reference. Amherst p.l.

STATON, W. B.: Postwar reference. Amherst p.l.

STONE, STEPHEN V.: Pvt. Enl. on 3/13/62 in Hanover Co. Reassigned to the Ashland Art. on 10/8/62 by S.O. 209, Headquarters, Army of Northern Virginia. Pres. on rolls from 10/31/62 until hospitalized in Richmond in 3-4/64. Admitted to Charlottesville Gen. Hosp. 4/27/64 with chronic dysentery. Died on 5/7/64.

STORY, W. P.: Postwar reference. Amherst p.l.

STRINGFELLOW, HENRY MARTYN: Jr. 1st Lt. B. Frederick Co. Elected to Sgt., 2/62. Assigned to battery 8/18/62. Transferred to Richardson's Battalion, 9/15/62. Age 22, 5'10", light comp., grey eyes, dark hair, student. NFR. Later ordnance officer.

STRINGFELLOW, ROBERT R.: 4th Sgt. B. Washington, D.C. Enl. 5/21/61, Beaver Dam Depot; age 18, student. Cpl. 8/5/61. 4th Sgt. (2/28/62). Discharged due to lung hemorrhage, 7/62; 6', dark comp., blue eyes, sandy hair. Drew veterans pension in Florida (#A04776).

SYDNOR, STEPHEN C.: Pvt. Enl. 3/15/62. Pres. Through (6/30/62). Transferred to Kirkpatrick's Co.

TAYLOR, EDMOND: 2nd Cpl. Enl. 3/25/62, Henrico Co. Elected 2nd Cpl. and absent sick (6/30/62). Transferred To Kirkpatrick's Co.

TERRELL, JOHN H.: Pvt. Enl. on 8/5/61 in Jamestown, James City Co. Pres. on 8/31/61 and 4/30/62 rolls. Reassigned to the Ashland Art. on 10/8/62 by S.O. 209, Headquarters, Army of N. Virginia. Pres. all later rolls thru 2/28/65. Paroled at Appomattox C.H., VA on 4/9/65. Res. of Hanover Co. Bro. of Joseph R., Nicholas, and William D.

TERRELL, JOSEPH R.: Pvt. Enl. on 3/14/62 in Hanover Co. Paid $50 bounty on 6/13/62. Reassigned to the Ashland Art. on 10/8/62 by S.O. 209, Headquarters, Army of Northern Virginia. KIA at Gettysburg, PA on 7/2/63, "had legs cut off at the hips."

TERRELL, LEWIS F.: 1st Lt. Enl. 5/21/61, Beaver Dam Depot; age 23, lawyer. Absent on leave (6/30/61). Appointed adjutant by Lt. Catesby Jones. 1st Lt. (2/28/62). Relieved 4/30/62.

TERRELL, NAPOLEON BONAPARTE: Pvt. B. Hanover Co. Enl. on 5/21/61 at Beaver Dam Depot, Hanover Co., VA, described as farmer, age 24, 5'7-", light comp., light hair, gray eyes. Furloughed 2/28/62. Pres. (4/30/62). In Chimborazo Hosp. #3 w/chronic diarrhea, 6/22/62. Sent to Lynchburg. Reassigned to the Ashland Art. on 10/8/62 by S.O. 209, Headquarters, Army of Northern Virginia. Pres. all rolls thru 2/28/65. Paroled at Ashland, VA on 4/21/65. Applied for pension in Hanover Co., VA in 1900, then age 64. Living in 1906.

TERRELL, NICHOLAS: 3rd Sgt. B. Hanover Co. Enl. 5/21/61, Beaver Dam Depot; age 19. (2/28/62). At Chimbo. #5 with fever, 6/10/62 - 6/15/62. Transferred To Woolfolk's Co.

TERRY, JAMES T.: Pvt. Enl. 5/21/61, Beaver Dam Depot. Described as a native of Hanover Co., laborer, age 22, 5'1-", dark comp., hazel eyes, black hair. Furloughed 2/28/62. Pres. (4/30/62). Admitted to Chimborazo Hosp. 6/10/62. RTD 6/16/62. Reassigned to the Ashland Art. on 10/8/62 by S.O. 209, Headquarters, Army of Northern Virginia. Pres. on all muster rolls thru 2/28/65. Paroled at Appomattox C.H., VA on 4/9/65. Res. of Hanover Co.

TERRY, JOHN FRANK: Pvt. Enl. 3/8/62 in Hanover Co., described as 24 years old, mechanic, native of Hanover Co., light hair, light comp., black eyes. Admitted to Petersburg, VA Hosp. with wrist problem. Furloughed on 9/5/62. Reassigned to the Ashland Art. on 10/8/62 by S.O. 209, Headquarters, Army of Northern Virginia. Pres. on all rolls thru 2/28/65. Paroled at Appomattox C.H., VA on 4/9/65. Applied for pension in Richmond in 1906. Entered Old Soldier's Home on 2/5/1907, age 67. Died on 5/23/1907 of old age.

TERRY, WILLIAM H.: Pvt. Enl. 5/15/62 in Hanover Co. Reassigned to the Ashland Art. on 10/8/62 by S.O. 209, Headquarters, Army of Northern Virginia. Pres. on all rolls thru 2/28/65. Paroled at Appomattox C.H., VA on 4/9/65. Res. of Hanover Co. Applied for pension in Hanover Co. about age 70. D. 8/19/1924, buried in the Taylorsville Baptist Church Cem., Spotsylvania C.H., VA.

THACKER, ANDREW JACKSON: Sgt., Enl. on 5/21/61 at Beaver Dam Depot, described as mechanic, age 24. Furloughed (2/28/62). Pres. (4/30/62). Reassigned to the Ashland Art. on 10/8/62 by S.O. 209, Headquarters, Army of Northern Virginia. Pres. on all rolls thru 2/28/65. Paroled at Ashland, VA on 4/21/65. Applied for pension in 1904 in Hanover Co., then age 67. Died on 1/11/1907 of apoplexy at 2215 Fairmont Ave., Richmond, VA. A record states that he had both hands amputated at the wrists, G.H. Petersburg, 8/1/62. Furloughed from hospital 9/5/62. D. 9/1917, Orange, VA.

THACKER, WILLIAM H.: Pvt. Enl. on 3/21/62 at Beaver Dam Depot, described as a 21 year old mechanic. Reassigned to the Ashland Art. on 10/8/62 by S.O. 209, Headquarters, Army of Northern Virginia. Absent sick on 10/31/62 roll. Pres. on all later rolls thru 2/28/65. Paroled at Ashland, VA on 4/21/65. Res. of Hanover Co. D. 10/1/1909 in Orange Co,. VA. Buried in family Cem.

THOMPSON, CHARLES R.: Pvt. Enl. on 5/21/61 in Hanover Co., described as a native of Louisa Co., farmer, age 28, 5'8", light comp., light hair, blue eyes. Furloughed (2/28/62). Pres. (4/62). Reassigned to the Ashland Art. on 10/8/62 by S.O. 209, Headquarters, Army of Northern Virginia. Pres. 10/31/62. Sick on 2/28/63 roll. Pres. on later rolls thru 4/30/64. Detailed to Pioneer Corps, Pickett's Division. Paroled at Ashland, VA on 5/6/65, occ. farmer. Applied for pension in 1902, age 69, in Hanover Co. D. 5/1/1920 in Hanover Co. of pneumonia and old age.

THOMPSON, JOHN LEWIS: Pvt. B. Louisa Co. Enl. on 5/21/61 in Hanover Co. Furloughed (2/28/62). Pres. (4/62). Reassigned to the Ashland Art. on 10/8/62 by S.O. 209, Headquarters, Army of Northern Virginia. WIA at Chancellorsville on 5/1-2/63, lost right arm. Retired from service on 3/16/64 by G.O. 34, Adjutant and Inspector General's Office. Extensive Hospitalizations. Transferred to Invalid Corps. on 1/11/65. Paroled at Beaver Dam, VA 5/8/65. D. 7/11/1920 at Beaver Dam, Hanover Co., age 82 of paralysis.

TREVILLIAN, JOHN: Pvt. Enl. 3/13/62. Transferred to Kirkpatrick's Co.

TUCKER, THOMAS G.: Postwar reference. Amherst p.l.

TURNER, BEVERLEY BRADSHAW: Pvt. Enl. 5/21/61, Beaver Dam Depot; age 19, student. Transferred 8/28/61 to Captain Douglas' Co. Lee Rangers, 9th Va. Cav.

TURNER, T. C.: Postwar reference. Died in service. Hanover p.l.

TYLER, RICHARD R.: Pvt. B. Hanover Co. Enl. 5/21/6, Beaver Dam Depot; age 22, mechanic. Absent sick (8/31/61). Furloughed (2/28/62); age 24, light comp., 5'7 1/4", blue eyes, dark hair, farmer. Pres. (4/62). D. 6/62.

WASHINGTON, WILLIAM H.: Pvt. Enl. ? Paroled 4/25/65.

WIATT, T.: Pvt. Enl.? Paroled 5/30/65, Campbell C.H.

WILLIAMSON, DABNEY: Pvt. Enl. 5/21/61, Beaver Dam Depot; age 20, farmer. Furloughed (2/28/62). Pres. (4/62). Transferred To Kirkpatrick's Co.

WILSON, W. A.: Ord. Sgt. Enl.? At Charlottesville G.H. w/fever 10/24/64 - 11/12/64. Paroled 5/4/65, Winchester.

WINSTON, ROBERT B.: Pvt. Enl. 3/2/62. Transferred To Kirkpatrick's Co.

WOOD, ALBERT: Pvt. Enl.? Paroled 5/18/65, Charlottesville, Va.

WRIGHT, GEORGE S.: Pvt. Enl. on 5/21/61 in Hanover Co. Reassigned to the Ashland Art. on 10/8/62 by S.O. 209, Headquarters, Army of Northern Virginia. Status not stated on 10/31/62 roll. On leave on

11/1/62. AWOL on 12/31/62 roll. Deserted at Millwood on 7/25/63. Pres. 4/30/64 thru 12/31/64 rolls. Transferred to Parker's Battery in exchange for E. S. Duffey in 2/65.

Magruder Light Artillery
Page's Battery

ALLEN, JAMES S.: Pvt. Enl. on 3/21/62 at Yorktown. Transferred to Bedford Art. on 10/1/62. Transferred to 4th Co. Washington Art. (Louisiana unit) on 11/17/62 by S.O. 242/9. Deserted, DFR 7/1/64. Res. of New Orleans, LA.

BARKER, BENJAMIN FRANKLIN: Pvt. Enl. 2/17/62, Jamestown. AWOL (6/30/62).

BELMER, HERMANN: Pvt. Enl. on 4/4/61 in New Orleans, LA in Zouave Battn. Joined Magruder Art. on 3/31/62 at Yorktown, VA. Transferred to Bedford Art. on 10/1/62. Sick on 6/63 roll and much of the rest of the war. Paroled at Appomattox. C.H., VA on 4/9/65. Res. of New Orleans, LA.

BISHOP, THOMAS: Pvt. Enl. 2/17/62, Jamestown. At Chimbo. #3, 4/6 - 4/14/62. Transferred to Smith's Bedford Artillery (hereafter B.A.)

BLEDSOE, ROBERT H.: 1st Sgt. Enl. 3/3/62, Richmond. WIA, 6/27/62. Orderly Sgt. (4/62). 1st Sgt. 6/30/62. Transferred to B.A.

BLUM, ERNEST: Sgt. Enl. 7/3/62, Yorktown. Transferred from Albemarle Artillery. Transferred to B.A.

BONHOMME, JEAN: Cpl. Enl. 1/27/62, Yorktown. AWOL (6/30/62).

BORMET, JOHN: Pvt. Enl. on 8/3/62 in Richmond, VA. Reassigned to the Bedford Art. on 10/8/62 by S.O. 209, Headquarters, Army of Northern Virginia. AWOL on all subsequent muster rolls.

BRADY, JOHN: Pvt. Enl. 1/27/62, Yorktown. AWOL (7/25/62).

BREMO, JOSEPH: Pvt. Enl. 1/27/62, Yorktown. AWOL (6/30/62).

BROWN, DIXON: Sgt. Enl. 5/29/61, Yorktown (Cosnahan's Battery). Sgt (4/30/62). Transferred to B.A.

BULMAN, FRANCIS W.: Pvt. Enl. on 1/27/62 at Yorktown. Reassigned to the Bedford Art. on 10/8/62 by S.O. 209, Headquarters, Army of Northern Virginia. Pres. until AWOL on 2/29/64 roll.

CALLIS, DANIEL: Pvt. Enl. 5/29/61, Yorktown (Cosnahan's Battery). Transferred to Page, 7/15/62. Transferred to B.A.

CAMILO, AUGUST: Pvt. Enl. 3/3/62, Yorktown. Transferred to B.A.

CARLETON, JOHN W.: Pvt. Enl. 1/27/62, Yorktown. Transferred to B.A.

CHAPMAN, WILLIAM P.: Pvt. Enl. on 7/15/61 in Cosnahan's Battery. Transferred to Magruder Art. on 7/15/61. Reassigned to Jordan's Bedford Art. on 10/8/62 by S.O. 209, HQ, ANV. Pres. or hospitalized most of the war. Pres. on 2/28/65 roll.

COBB, RICHARD H.: Pvt. Enl. on 3/31/62 at Yorktown. Reassigned to Jordan's Bedford Art. on 10/8/62 by S.O. 209, HQ, ANV. Pres. or in hosp. remainder of the war. Paroled at Mechanicsville, King William Co., VA on 5/2/65. Living in Blendon Dist., Nottoway Co., VA in 1870, then age 35.

CURL, WILLIAM E.: Pvt. Enl. on 2/17/62 at Jamestown, VA. Reassigned to Jordan's Bedford Art. on 10/8/62 by S.O. 209, HQ, ANV. Pres. on all rolls thru 2/28/65. Paroled at Appomattox C.H., VA on 4/9/65. Res. of Charles City Co.

DEW, R. H.: Pvt. Enl. 1/27/62, Yorktown. Present through 6/30/62. NFR.

DILLON, FRANCIS: Pvt. Enl. 1/27/62, Yorktown. NFR.

DRISCOLL, DANIEL J.: Cpl. Enl. 3/27/61 in New Orleans, LA, age 28, ambrotypist, single, res. of New Orleans, LA. Joined the Magruder Art. on 3/3/62 at Yorktown, VA. Cpl. (6/30/62). Reassigned to Jordan's Bedford Art. on 10/8/62 by S.O. 209, HQ, ANV. Pres. until he Transferred to Co. 2 Washington Art. LA Vols. on 7/16/64. Paroled at Lynchburg, VA on 4/19/65.

FLICK, GEORGE: Pvt. Enl. 3/31/62, Yorktown. AWOL (6/30/62).

FLYNN, ____: Pvt. Enl. 3/31/62, Yorktown. AWOL (6/30/62).

GOODE, JOSEPH H.: Pvt. Enl. on 2/17/62 at Jamestown, VA. Reassigned to Jordan's Bedford Art. on 10/8/62 by S.O. 209, HQ, ANV. Pres. or hospitalized until he died on 2/7/64 of variola at Howard's Grove Hosp., Richmond, VA.

GRAVES, JOHN: Pvt. Enl. 5/29/61, Yorktown (Cosnahan's Co.). Transferred from Cosnahan's 7/25/62. Transferred to B.A.

GREEN, PETER: Pvt. Enl. on 5/29/61 at Yorktown, VA in Cosnahan's Battery. Transferred to Magruder Art. on 7/15/62. Reassigned to Jordan's Bedford Art. on 10/8/62 by S.O. 209, HQ, ANV. WIA at Gettysburg, PA on 7/2/63, shell would in left thigh. Hospitalized most of the remainder of the war. Age 36, living in Grafton Dist., York Co., VA in 1870.

HALL, SMITH: Pvt. Enl. on 5/18/61 at Upper Grafton in 32nd VA Inf. Transferred to Cosnahan's Battery and to the Magruder Art. on 7/15/62. Reassigned to Jordan's Bedford Art. on 10/8/62 by S.O. 209, HQ, ANV. Deserted at Chester Gap 7/21-28/63, took oath at the Old Capitol Prison, Washington, DC. Res. of York Co., VA, 5'4", dark comp., black hair, black eyes, sent north.

HASWELL, GEORGE WASHINGTON: Pvt. See 52nd Va. Inf. Enl. 2/17/62, Jamestown. AWOL (6/30/62). Deserter, walked into Federal lines at Williamsburg, 2/65; dark comp., 5'7½", brown hair, grey eyes, resident of New Kent Co. Sent to Ft. Monroe and took oath 2/28/65.

HERRING, B.: Cpl. Enl. 1/27/62, Yorktown. AWOL (6/30/62). NFR.

HOGAN, TIMOTHY: Pvt. Enl. on 2/14/62 in Richmond, VA in Capt. S. P. Hamilton's GA Battery. Transferred to Magruder Art. on 7/23/62. Reassigned to Jordan's Bedford Art. on 10/8/62 by S.O. 209, HQ, ANV. Pres. or hospitalized most of the remainder of the war. Paroled at Lynchburg, VA on 4/15/65.

HOGGE, ADDISON RILEY: Pvt. Enl. on 5/29/61 at Yorktown in Cosnahan's Battery. Transferred to Magruder Art. on 7/15/62. Reassigned to Jordan's Bedford Art. on 10/8/62 by S.O. 209, HQ, ANV. Extensive hospitalization, final status however, Pres. on 2/28/65 roll.

HOGGE, JOHN W.: Pvt. Enl. on 5/29/61 at Yorktown in Cosnahan's Battery. Transferred to Magruder Art. on 7/15/62. Reassigned to Jordan's Bedford Art. on 10/8/62 by S.O. 209, HQ, ANV. Extensive hospitalization, final status show as deserter on 2/28/65 roll.

HOTTINGER, S.: Pvt. Enl. 3/31/62, Yorktown. AWOL (6/30/62). NFR.

HOWARD, THOMAS C.: Sgt. Enl. on 5/29/61 at Yorktown in Cosnahan's Battery. Transferred to Magruder Art. as Sgt. on 7/15/62. Pres. most of the war. Paroled at Appomattox C.H., VA on 4/9/65. Res. of York Co., VA.

HURT, JOSEPH S.: Pvt. Enl. on 1/27/62 at Yorktown, VA. Reassigned to Jordan's Bedford Art. on 10/8/62 by S.O. 209, HQ, ANV. Court-martialed 4/16/63. WIA at Chancellorsville on 5/1- 4/63. Deserted, went into Federal lines, received at Fort Monroe on 4/5/64. Took oath, sent to NY. Res. of King & Queen Co., 5'10 3/4", dark comp., black hair, grey eyes.

JONES, HENRY F.: Sgt. Enl. on 5/29/61 at Yorktown at Cosnahan's Battery. Transferred to Magruder Art. as Sgt. on 7/15/62. Reassigned to Jordan's Bedford Art. on 10/8/62 by S.O. 209, HQ, ANV. Appointed Sgt. in Bedford Art. on 3/1/63. Pres. all rolls. Paroled at Appomattox C.H., VA on 4/9/65. Res. of Warwick Co., VA, living there in 1870, then age 31.

JONES, JAMES L.: Pvt. Enl. 3/25/62, Lynchburg bu Lt. Taylor. Transferred from Hamilton's Ga. Battery 7/23/62. Transferred To B.A.

KEIFER, ____: Pvt. Enl. 3/31/62, Yorktown. AWOL (6/30/62). NFR.

KEMPLETON, EADON or EDWARD: Pvt. Enl. on 3/25/62 at Lynchburg, VA in Capt. S. P. Hamilton's GA Battery. Transferred to Magruder Art. on 7/23/62. Reassigned to Jordan's Bedford Art. on 10/8/62 by S.O. 209, HQ, ANV. Pres. or accounted for thru 2/28/65.

LAFORGE, ____: Pvt. Enl. 4/10/62, Yorktown. AWOL (7/25/62). NFR.

MAGRUDER, GEORGE ALLEN: 1st Lt. Present through (7/25/62). NFR.

McBRIDE, ____: Pvt. Enl. 4/10/62, Yorktown. AWOL (6/30/62).

McKAY, JOHN: Sgt. Enl. 1/27/62, Yorktown. Present through (7/25/62). NFR.

MELOTH, CHRISTIAN: Pvt. Enl. on 4/18/61 at New Orleans, LA in Zouave Battn. Transferred to Magruder Art. on 3/31/62 at Yorktown, VA. Reassigned to Jordan's Bedford Art. on 10/8/62 by S.O. 209, HQ, ANV. Paroled on 4/10/65 at Appomattox C.H., VA. Res. of New Orleans, LA, age 44, 5'10", sandy comp., dark hair, dark eyes.

MEYER, CORT: Pvt. Enl. on 4/4/61 in New Orleans, LA. Transferred to Magruder Art. on 3/31/62 at Yorktown. Reassigned to Jordan's Bedford Art. on 10/8/62 by S.O. 209, HQ, ANV. Hospitalized extensively, however, paroled at Appomattox C.H., VA on 4/9/65. Res. of New Orleans, LA.

MILES, RICHARD: Cpl. Enl. 1/27/62, Yorktown. AWOL (6/30/62). Present (7/25/62). NFR.

MILLIRON, ALEXANDER R.: Pvt. Enl. 3/25/62 in Lynchburg in Capt. S. P. Hamilton's GA Battery. Transferred to Magruder Art. on 7/23/62. Reassigned to Jordan's Bedford Art. on 10/8/62 by S.O. 209, HQ, ANV. MWIA in thigh, which was amputated, at Chancellorsville, VA 5/1- 4/63. Died on wounds at Howard's Grove Hosp., Richmond, VA on 8/4/63.

MILLS, CHARLES W.: Pvt. Enl. on 3/25/62 in Capt. S. P. Hamilton's GA Battery. Transferred to Magruder Art. on 7/23/62. Reassigned to Jordan's Bedford Art. on 10/8/62 by S.O. 209, HQ, ANV. Sick from 6/63 roll thru mid-1864. Pres. on 10/31/64 and 2/28/65 rolls. Paroled at Appomattox C.H., VA. Res. of Franklin Co., VA.

MITCHELL, CHARLES: Pvt. Enl. 2/18/62, Yorktown. Absent in hospital (6/30/62).

MOORE, JAMES ELLETT: Pvt. Enl. on 1/27/62 at Yorktown, VA. Reassigned to Jordan's Bedford Art. on 10/8/62 by S.O. 209, HQ, ANV. AWOL 2/29/64 roll. Pres. on 10/31/64 and 2/28/65 rolls.

MOORE, WILLIAM R.: Cpl. Enl. 1/27/62, Yorktown. Cpl. (6/30/62). At Winder Hospital 9/25/62 - 1/63.

MOIRIZOLI, BERNARD: Pvt. Enl. on 3/31/61 in New Orleans, LA, born at Tacino, LA, age 24, tailor, res. of New Orleans, LA. Transferred to Magruder Art. on 3/31/62 at Yorktown, VA. Reassigned to Jordan's Bedford Art. on 10/8/62 by S.O. 209, HQ, ANV. Transferred to 4th Co., Washington LA Art. on 11/17/62. Pres. on all rolls thru 2/28/65.

MURPHY, ______: Pvt. Enl. 4/10/62. AWOL (6/30/62).

MURRAY, WILLIAM C.: Pvt. Enl. 3/15/62. AWOL (6/30/62).

NEWMAN, THOMAS: Cpl. Enl. on 7/17/62 in Richmond, VA. Reassigned to Jordan's Bedford Art. on 10/8/62 by S.O. 209, HQ, ANV. Taken POW in King & Queen Co. on 3/11/64, sent to Fort Monroe and then to Point Lookout, MD. Exchanged. Pres. on 12/31/64 and 2/28/65 rolls. Paroled at Appomattox C.H., VA on 4/9/65. Res. of Warwick Co.

OLIVER, JUAN: Pvt. Enl. 4/10/62, Yorktown. AWOL (6/30/62).

PAGE, THOMAS JEFFERSON: Captain. Only name slip in unit's CSR.

PITZENBERGER, VINCENT: Bugler. Enl. on 5/2/61 in New Orleans, LA in Co. F, Zouave LA Battn. Transferred to Magruder Art. on 3/31/62. Reassigned to Jordan's Bedford Art. on 10/8/62 by S.O. 209, HQ, ANV. Died of acute dysentery on 7/29/64.

PROUT, JOHN: Pvt. Enl. 1/27/62, Yorktown. AWOL (6/30/62).

REILER, CHARLES: Pvt. Enl. on 3/26/61 in New Orleans, LA in Co. A, Zouave LA Battn. Transferred to Magruder Art. on 3/31/62. Reassigned to Jordan's Bedford Art. on 10/8/62 by S.O. 209, HQ, ANV. WIA at Gettysburg, PA on 7/3/63, wounded in jaw impairing motion. Pres. 10/31/64 thru 2/28/65 detailed as saddler in field part. Taken POW at Farmville, VA on 4/6/65, sent to Newport News where held until released on oath on 6/16/65. Born in Germany, res. of New Orleans, LA, saddler, age 32, 5'8", light comp., light hair, blue eyes.

SCHNEIDER, FREDERICK: Pvt. Enl. 4/10/62, Yorktown. KIA 6/27/62.

SCHWEIZER, ______: Pvt. Enl. 4/10/62. AWOL (6/30/62).

SHERWELL, WILLIAM: Cpl. Enl. 7/25/62, Richmond. Substitute for William Smith. Cpl. through 7/15/62. Transferred to B.A.

SMITH, G. N.: Pvt. Enl. 1/27/62, Yorktown. Transferred to ironclad steamer, Virginia, 2/9/62.

SMITH, GEORGE W.: Pvt. Enl. on 5/29/61 at Yorktown in Cosnahan's Battery. Transferred to Magruder Art. on 7/15/62. Reassigned to Jordan's Bedford Art. on 10/8/62 by S.O. 209, HQ, ANV. MWIA at Gettysburg, PA, died on 7/4/63 per one record, another indicates that he was admitted to Chimborazo Hosp. #1 on 6/11/64 with rubeola erysipelas.

SMITH, JOHN DONNELL: 1st Lt., Enl. on 1/62 as aide-de-camp to Gen. Magruder. Assigned to Magruder Art. at unknown date. Reassigned to Jordan's Bedford Art. on 10/8/62 by S.O. 209, HQ, ANV. as 1st Lt. Promoted to Capt. by 10/31/64. Paroled at Appomattox C.H., VA on 4/9/65. Paroled as a res. of Baltimore, MD B. 6/5/1829 in Baltimore, MD. Graduated with B.A. degree from Yale University in 1847 and from Georgetown Univ. in 1857, admitted to the bar, but never practiced law. Traveled extensively in Europe. D. 12/2/1928, age 99- years in Baltimore, MD.

SOMMER, JOHN: Pvt. Enl. on 3/31/62 at Yorktown. Reassigned to Jordan's Bedford Art. on 10/8/62 by S.O. 209, HQ, ANV. Discharged on 2/3/63.

STANSBURG, CHARLES: Pvt. Enl. 3/7/62. Present through (10/28/62). Transferred to B.A.

STRASSER, JACOB: Pvt. Enl. on 3/29/61 in New Orleans, LA in Co. A, Zouave LA Battn. Transferred to Magruder Art. on 1/27/62 at Yorktown. In Danville Hosp. 5/5/62 with gunshot wounds. RTD 6/3/62. Reassigned to Jordan's Bedford Art. on 10/8/62 by S.O. 209, HQ, ANV. Pres. thru 2/28/65 roll. Res. of New Orleans, LA.

SWARTZ, GEORGE W.: Pvt. Enl. on 5/29/61 at Yorktown in Cosnahan's battery. Transferred to Magruder Art. on 7/15/62. Reassigned to Jordan's Bedford Art. on 10/8/62 by S.O. 209, HQ, ANV. Court Martialed, Gen. Orders. #57, HQ, ANV 4/18/63. Deserted, captured at Chester Gap on 7/21/63, sent to Old Capitol Prison, Washington, DC where released on oath and sent north. Res. of York Co., 5'5", dark comp., brown hair, hazel eyes.

THOMAS, WILLIAM: Pvt. Enl. on 5/29/61 at Yorktown, VA in Cosnahan's Battery. Transferred to Magruder Art. on 7/15/62. Reassigned to Jordan's Bedford Art. on 10/8/62 by S.O. 209, HQ, ANV. In arrest on 2/28/63 roll thru 11/12/63. AWOL on 2/29/64 roll.

TWIGGS, RICHARD T.: Pvt. Enl. 1/27/62, Yorktown. AWOL (6/30/62).

WARNER, JOHN: Pvt. Enl. 1/27/62, Yorktown. AWOL (6/30/62). NFR.

WEST, YANCEY: Pvt. Enl. ? Paroled 4/24/65, Ashland.

WRIGHT, JAMES G.: Pvt. Enl. on 3/25/62 at Lynchburg, VA in Capt. S. P. Hamilton's GA Battery. Pres. on 4/30/62 and 6/30/63 rolls. Reassigned to Jordan's Bedford Art. on 10/8/62 by S.O. 209, HQ, ANV. Deserted during 9-10/62.

WRIGHT, JOHN: Pvt. Enl. on 5/29/61 at Yorktown in Cosnahan's Battery. Transferred to Magruder Art. on 7/15/62. Reassigned to Jordan's Bedford Art. on 10/8/62 by S.O. 209, HQ, ANV. Pres. most of the war. Paroled at Appomattox C.H., VA on 4/9/65. Res. of York Co., VA.

YANCEY, STEPHEN D.: 2nd Lt. Only one slip in CSR.

ABBOTT, JAMES T.: 2nd Lt. See 16th Va. Inf. Enl. 5/1/61 at Manchester, Chesterfield Co., VA. Promoted to 5th Sgt. in 11/61. Reduced to ranks, 6/4/62. 3rd Lt. 5/1/62. Assigned to duty 7/15/62. Pres. On detached service 12/30/62. 2nd Lt. (2/28/62). Transferred to Courtney's Art.

ABBOTT, JOHN H.: Pvt. See 16th Va. Inf. Enl. 5/1/61 at Manchester. Absent sick 10/16/62 - (2/28/62). Wounded in hand (10/31/62). Alive on 9/20/1902. Transferred to Courtney's Art.

ANDERSON, JAMES B.: Pvt. Enl. 5/1/61 at Manchester.

ANDERSON, JOHN B.: Pvt. See 16th Va. Inf. Enl. 5/1/61 at Manchester. AWOL 8/8/62. At Chimbo., 10/22/62. At Farmer's Hotel, Manchester. To Chimbo. #5 with debility 10/14/63 and Transferred To private quarters, 10/24/63. Transferred to Courtney Art. Deserted 2/20/64.

ATKINSON, GEORGE A.: Pvt. B. Chesterfield Co. See 16th Va. Inf. Enl. 7/28/61 at Sewell's Point, near Norfolk. Absent sick 6/15/62. AWOL. Pres. (1/1/63) - (2/28). Age 26, fair comp., blue eyes, brown hair, carpenter. Transferred to Courtney Art.

BASS, MARCELLUS CLAIBORNE: Pvt. See 16th Va. Inf. Enl. 5/1/61 at Manchester. Pres. through (2/28/63). Transferred to Courtney Art.

BLANKENSHIP, WILLIAM B.: Pvt. B. Chesterfield Co. See 16th Va. Inf. Enl. 5/1/61 at Manchester. AWOL 7/15/62. Discharged on 8/1/62; age 39, 5'11½", dark comp., blue eyes, brown hair, laborer.

BLANTON, JOHN A.: Pvt. See 16th Va. Inf. Enl. 5/1/61 at Manchester. Abs. Sick (7/24/62). Pres. (11/27/62) - (2/28/63). Transferred to Courtney Art.

BLANTON, JOSEPH J.: Pvt. See 16th Va. Inf. Enl. on 5/14/61 at Sewell's Point, near Norfolk, VA. Abs. Sick 7/15/62. Pres. (1/1/63) - (2/28). Transferred to Courtney Art.

BLUNT, WILLIAM ROBERT: Pvt. B. Henrico Co. See 16th Va. Inf. Enl. on 5/1/61 at Manchester. Abs. Sick 7/4/62 - (2/28/63). Age 19, 5'8", fair comp., blue eyes, light hair, brick maker. Transferred to Courtney Art.

BOWMAN, ROBERT C.: 4th Cpl. See 16th Va. Inf. Enl. 8/15/61 at Sewell's Point. Promoted to Cpl., 12/1/61. 4th Cpl. (7/24/62). Transferred to Courtney Art. Surrendered at Appomattox.

BOWMAN, WILLIAM A. D.: 2nd Cpl. See 16th Va. Inf. Enl. 5/1/61 at Manchester. 2nd Cpl. 1/1/63. Transferred to Courtney Art. Paroled 4/29/65, Manchester.

BROWDER, PHILIP B.: Pvt. See 16th Va. Inf. Enl. 5/1/61 at Manchester. Pres. Through (2/28/63). Transferred to Courtney Art.

BROWN, BENJAMIN M.: Pvt. See 16th Va. Inf. Enl. 5/1/61 at Manchester. AWOL 8/62. Pres. Through (2/28/63). Transferred to Courtney Art.

BRYANT, GEORGE F.: Pvt. See 16th Va. Inf. Enl. 5/1/61 at Manchester. Detached in Richmond, 5/15/62. Pres. Through (2/28/63. Transferred to Courtney Art.

BUCHANAN, JAMES: Pvt. See 16th Va. Inf. B. 4/6/41, Glasgow, Scotland. Enl. 5/1//61 in Manchester. AWOL 7/4/62 - 1/10/63. Transferred to Courtney Art.

BURNETT, SAMUEL E.: Pvt. See 16th Va. Inf. Enl. 5/1/61 at Manchester. Transferred to Otey's Co., 7/15/62.

CHEATHAM, ALFRED B.: Pvt. See 16th Va. Inf. Enl. 5/1/61 at Manchester. Present through transfer to Courtney Art.

CHEATHAM, JAMES W.: 4th Cpl. See 16th Va. Inf. Enl. 6/26/61 at Sewell's Point, near Norfolk. AWOL 12/2/62. Pvt. 1/63. Present through transfer to Courtney Art.

CLARKE, EDWARD S.: Pvt. See 16th Va. Inf. Enl. 5/14/61, Sewell's Point. AWOL 5/7/62 - (2/28/63). Transferred to Courtney Art.

CLARKE, JAMES A.: Pvt. See 16th Va. Inf. Enl. 6/26/61, Sewell's Point. AWOL 5/7/62 - (2/28/63. Transferred to Courtney Art.

CLARKE, SHIRLEY H.: Pvt. See 16th Va. Inf. Enl. 5/1/61 at Manchester. Abs. sick 6/29/62 - (2/28/63). Transferred to Courtney Art.

COX, JOSEPH EDWIN: 3rd Lt. B. 9/27/1837. University of Virginia student. See 16th Va. Inf. Enl. 5/1/61 at Manchester. Died on 12/26/1861.

COX, THOMAS J.: Pvt. See 16th Va. Inf. Enl. 5/1/61 at Manchester. KIA 6/8/62 at Cross Keys.

CROSS, JOHN W.: Pvt. B. Chesterfield Co. See 16th Va. Inf. Enl. 5/1/61 at Manchester. Absent sick 6/29/62 - (2/28/63). Age 26, 5'9", hair comp., grey eyes, brown hair, blacksmith. Transferred to Courtney Art.

CROSTICK, JOHN H.: Pvt. B. Chesterfield Co. See 16th Va. Inf. Enl. 5/22/61 at Sewell's Point near Norfolk. Absent sick 6/29/62 - (2/28/63). Age 31, 5'10", dark comp., grey eyes, black hair, painter. Transferred to Courtney Art.

CROSTICK, THOMAS O.: Pvt. See 16th Va. Inf. Enl. 7/24/61 at Sewell's Point near Norfolk. Wagoner, 7/10/62. Teamster through (2/28/63). Transferred to Courtney Art.

DAVIS, PETERSON S.: Pvt. B. Greenville, Va. See 16th Va. Inf. Enl. 5/1/61 at Manchester. Discharged on surgeon's cert. 2/20/63.; age 24, 5'8", light comp., grey eyes, light hair, shoemaker. See 6th Va. Inf.

DUNAVANT, ARCHER: Pvt. See 16th Va. Inf. Enl. 5/1/61, Manchester. Furloughed 2/20/63. Rejoined 16th Va. Inf. Surrendered at Appomattox.

DUNAWAY, HENRY H.: Pvt. See 16th Va. Inf. Enl. on 7/24/61 at Sewell's Point. Deserted 6/15/62 - 1/4/63. Transferred to Courtney Art.

EMERY, EDWARD: Pvt. B. Surrey Co. See 16th Va. Inf. Enl. 5/14/61 at Sewell's Point. AWOL 5/8/62. Discharged "over 35", 9/19/62; age 42, 5'4", fair comp., light blue eyes, sandy hair, wool spinner.

ENGLAND, ROBERT J.: Pvt. See 16th Va. Inf. Enl. 5/14/61 at Sewell's Point near Norfolk. Deserted 6/15/62 - 1/13/63. Transferred to Courtney Art.

ENGLISH, WILLIAM L.: Pvt. See 16th Va. Inf. Enl. 4/11/62 at Sewell's Point. Drafted from militia and assigned 4/62 - 2/2/63. Transferred. to Courtney Art.

EPPES, GEORGE H.: Pvt. See 16th Va. Inf. Enl. 5/1/61 at Manchester. Described as being born in Dinwiddie Co., occ. weaver, 5'8", dark comp., hazel eyes. Transferred out of company on 9/22/61. Listed as wagoner 5/15/62. Absent sick 7/10/62. Present 2/28/63. Transferred to Courtney Art.

EUBANKS, NATHAN A.: 2nd Lt. Enl. 5/1/61 at Manchester. Furloughed on 12/62. Relieved 7/4/62.

FARMER, FESTUS C. D.: Pvt. Enl. on 5/1/61 at Manchester. Detached on 10/5/61 in Richmond.

FOSTER, JOHN H.: QM Sgt. Enl.? 5th Sgt (7/24/62). Absent sick in hosp., Lynchburg, 2/11/63. QM Sgt. 1/1/63.

FRITH, F. or T. D.: Pvt. Enl. on 4/11/62 at Franklin Co. (Drafted/conscript). AWOL 4/11/62.

FURCRON, JAMES M.: Pvt. See 16th Va. Inf. Enl. 5/1/61 at Manchester. AWOL 5/8/62.

FURCRON, JOHN T.: Pvt. See 16th Va. Inf. Enl. 5/14/62, Sewell's Point. AWOL 12/24/62 - 12/30/62. Transferred to Courtney Art.

FURCRON, JUNIUS A.: Pvt. Enl. 1/15/63, Caroline Co. Transferred to Courtney Art. Detailed per S.O. #212/27, 9/7/63.

GARNETT, LELAND B. S.: Pvt. See 16th Va. Inf. Enl. 7/9/61 at Sewell's Point. AWOL 5/8/62.

GIBBS, THOMAS E.: 1st Lt. See 16th Va. Inf. Enl. 5/1/61, Manchester. Relieved 9/1/62. Commanding 9/19/62.

GILL, CHASTINE: Pvt. See 16th Va. Inf. Enl. 5/1/61 at Manchester. PWR only.

GILL, LUTINUS W.: Pvt. See 16th Va. Inf. Enl. 5/1/61, Manchester. Absent sick 5/8/62. Died 7/30/62.

GOLDSTEIN, JACOB: Pvt. See 16th Va. Inf. Enl. 4/11/62, Franklin Co. (Conscript). AWOL 4/11/62. Served in militia.

GRUBBS, RICHARD D.: Pvt. See 16th Va. Inf. Enl. 5/1/61, Manchester. At Chimbo #3 w/dysentery, 7/17/62 - 7/29/62. Transferred to Co. E, 44th Va. Inf. Per S.O. #237/7.

HALL, JOHN W.: 3rd Cpl. See 16th Va. Inf. Enl. 3/21/62, Sewell's Point. 3rd Cpl. 1/63.

HANOVER, HENRY B.: Pvt. Enl. 2/2/63, Caroline Co. Substitute for John R. Zeigler. Transferred to Courtney Art.

HARRIS, JOHN W.: Pvt. See 16th Va Inf. Enl. 5/1/61, Manchester. Present through to transfer to Courtney Art.

HEFFLEY, JOSEPH: Pvt. See 16th Va. Inf. Enl. 5/1/61 at Manchester. Transferred to 39th Battalion Va. Cav., 10/62.

HODGES, HEZEKIAH: Pvt. See 16th Va. Inf. Enl. 4/11/62, Franklin Co., as a conscript. AWOL 4/11/62. D. 5/62, Danville, Va.

HODGES, JAMES T.: Pvt. See 16th Va. Inf. Enl. 4/11/62, Franklin Co., as a conscript. Present through transfer to Courtney Art.

HODGES, SAMUEL or LEMUEL: Pvt. See 16th Va. Inf. Enl. 4/11/62, Franklin Co., as a conscript. AWOL 4/11/62. NFR.

HOLCOMB, JAMES or JOSEPH L.: Pvt. Pvt. Enl. 4/11/62, Franklin Co., conscript. Present through transfer to Courtney Art. Alive in 1909. Res. of Franklin Co.

HOLDER, MATHEW T.: Pvt. See 16th Va. Inf. Enl. 5/14/62, Sewell's Point. Present through to transfer to Courtney Art.

HOLT, WILLIAM H.: 1st Cpl. Se 16th Va. Inf. Enl. 5/1/61, Manchester. 1st Cpl. 5/14/62. 3rd Cpl. (11/27/62). Pvt. 11/22/62. Absent sick 12/26/62.

HUBAND, CORNELIUS: Pvt. B. Chesterfield Co. See 16th Va. Inf. Enl. 5/1/61 at Manchester. WIA in arm at Cross Keys, 6/8/62. Age 25, 5'3", fair comp., blue eyes, black hair, laborer.

HUBAND, HENRY: Pvt. See 16th Va. Inf. Enl. 5/1/61 at Manchester. Present through transfer to Courtney Art.

HUGHSTON, WILLIAM H.: Pvt. See 16th Va. Inf. Enl. 4/11/62, Franklin Co., conscript. Absent sick 7/4/62. NFR.

HUTCHINSON, WILLIAM HENRY: Bugler. B. 1826. See 16th Va. Inf. Absent sick 6/12/62. Pres. (1/1/63). Discharged, provided substitute (David Joice), 2/13/63. See Courtney Artillery. D. after 8/19/1921.

JARVIS, FRANCIS: Pvt. See 16th Va. Inf. Enl. 5/1/61 at Manchester. AWOL 5/8/62.

JENKINS, WILLIS J.: Pvt. See 16th Va. Inf. Enl. 4/11/62, Franklin Co., conscript. Deserted 5/15/62.

JETER, JAMES A.: Pvt. See 16th Va. Inf. Enl. 5/1/61 at Manchester. Present through transfer to Courtney Art.

JIMMERSON, ZACHARIAH: Pvt. See 16th Va. Inf. Enl. 7/28/61 at Sewell's Point. AWOL 7/4/62.

JOICE, DAVID: Pvt. Enl. 2/12/63, Caroline Co. Substitute for William R. Hutchinson. Present through transfer to Courtney Art.

JORDAN, DEWITT CLINTON: Pvt. See 16th Va. Inf. Enl. 5/1/61 at Manchester. At Richmond on detached service 11/61 per S.O. #199/2, 10/31/61. Deserter per L.R. #506, 8/20/63.

KAHN, ISAAC: Pvt. See 16th Va. Inf. Enl. 5/1/61 at Manchester. Discharged 5/4/62. Claimed French protection. See also 6th Va. Inf.

KELLY, ANDREW: Pvt. See 16th Va. Inf. Enl. 5/1/61 at Manchester. AWOL 8/1/62. See also 6th Va. Inf.

KELLY, JAMES M.: Pvt. See 16th Va. Inf. Enl. 5/1/61, Manchester. Wagoner, 5/15/62.

KELLY, SPENCER: Musician. Enl. 7/19/61 at Sewell's Point. Discharged 5/10/62.

KESLER, PETER: Pvt. See 16th Va. Inf. Enl. 4/11/62 at Franklin Co., conscript. AWOL since sent to soldiers home, 5/31/62. Furloughed 6/7/62.

LANNAN, MICHAEL J.: Pvt. See 16th Va. Inf. Enl. 5/1/61 at Manchester. AWOL, 5/8/62.

LEONARD, RICHARD T.: Pvt. See 16th Va. Inf. Enl. 5/1/61 at Manchester. Absent sick 6/13/62. Present (1/1/63). Absent sick in Lynchburg hospital 2/63. NFR.

LOVE, WILEY F.: Pvt. See 16th Va. Inf. Enl. 4/11/62 at Franklin Co., conscript. Previously served in militia. Absent sick, 8/1/62.

LOWRY, ROBERT P.: Pvt. B. Goochland Co. See 16th Va. Inf. Enl. 5/1/61 at Manchester. Absent sick 5/1/62. Discharged on 9/11/62 for intermittent fever with pneumonia and enlargement of the spleen; shoemaker, 5'8", fair comp., blue eyes, brown hair.

MACK, JAMES WILLIAM: Pvt. B. 5/1828. See 16th Va. Inf. Enl. on 5/14/61 at Sewell's Point. Absent detached in Richmond 11/20/62 - at least 2/28/63. Transferred to the Courtney Art. Res. of Richmond, VA. D. 7/24/1909 in Richmond, VA.

MADDOX, HERMAN L.: Pvt. See 16th Va. Inf. Enl. 5/1/61, Manchester. AWOL 10/20/62.

MASON, HOWARD: Pvt. See 16th Va. Inf. Enl. 5/1/61 at Manchester. Absent 6/20/62. Present (2/28/63). Discharged on 9/22/62; age 19, 6', florid comp., black eyes.

McGEE, CHARLES W.: Pvt. See 16th Va. Inf. Enl. 5/1/61 at Manchester. Present through transfer to Courtney Art.

MONTGOMERY, BENJAMIN D.: 2nd Sgt. B. Mecklenburg Co. See 16th Va. Inf. Enl. 5/1/61 at Manchester. Discharged 7/15/62; age 36, 5'2½", light comp., grey eyes, light hair, painter.

MOODY, BENJAMIN A.: Pvt. B. Chesterfield Co. See 16th Va. Inf. Enl. 5/1/61 at Manchester. AWOL 7/4/62. Discharged 8/1/62; age 36, 6'2½", dark comp., blue eyes, grey hair, blacksmith.

MURDEN, GREEN H.: Pvt. See 16th Va. Inf. Enl. 5/1/61 at Manchester.

PALMER, JOHN: Pvt. Enl. 4/11/62 at Huger Barracks near Norfolk, conscript. Previous service in militia. Resident of Chesterfield Co. Deserted 5/20/62. Transferred to Co. B, 39th Va. Cav. Paroled 4/27/65, Manchester.

PERRY, ARTEMUS H.: 2nd Lt. See 16th Va. Inf. Enl. 5/1/61 at Manchester. Promoted on 10/4/61 to 4th Sgt. Pres. 4/30/1862 roll. Absent sick 7/7/62. Lt. 5/1/62. Assigned 7/15/62. Resigned 3/21/63. Resignation accepted 4/2/63.

PERRY, CYRENIUS T.: 3rd Sgt. See 16th Va. Inf. Enl. 5/1/61 at Manchester. Transferred to Courtney Art.

POND, ATLAS T.: Pvt. B. Sussex Co. See 16th Va. Inf. Enl. 5/14/61 at Sewell's Point. Described as 6'4-", fair comp., light hair, blue eyes. MWIA 12/13/62, battle of Fredericksburg. Died on 1/22/63. Effects recd. 2/17/63, cert# 547.

POND, MAGNIS W.: Pvt. See 16th Va. Inf. Enl. 5/1/61 at Manchester. Transferred to Courtney Art.

ROUNDTREE, ALEXANDER W.: 3rd Sgt. B. Petersburg, Va. See 16th Va. Inf. Enl. 5/1/61 at Manchester. Discharged as 3rd Sgt, 7/15/62; age 17, 5'4", light comp., blue eyes, red hair, laborer.

RUSHBROOK, JOSEPH W.: Pvt. See 16th Va. Inf. Enl. 5/11/61 at Sewell's Point. Absent sick 6/20/62. Present (1/1/63). Transferred to Courtney Art.

SCARBOROUGH, JOHN C.: Pvt. See 16th Va. Inf. Enl. 5/1/61, Manchester. Absent sick 6/20/62 Present through transfer to Courtney Art.

SCOTT, ROGER: Pvt. See 16th Va. Inf. Enl. 5/1/61 at Manchester.

SHELL, JOHN: Pvt. See 16th Va. Inf. Enl. on 5/1/61 at Manchester. KIA at Cross Keys, 6/8/62.

STANLEY, PAUL: Pvt. See 16th Va. Inf. Enl. 5/1/61 at Manchester.

STONE, AUGUSTINE T.: Pvt. See 16th Va. Inf. Enl. 5/1/61 at Manchester. AWOL 7/15/62.

TENNILLE, JAMES T.: Pvt. See 16th Va. Inf. Enl. 5/1/61 at Manchester. Absent in Richmond, detailed 11/1/61.

TIMBERLAKE, CHARLES M.: Pvt. See 16th Va. Inf. Enl. 5/1/61 at Manchester. Detached as artificer, 9/1/62. AWOL 12/24/62 - 12/30/62. Transferred to Courtney Art.

TINSLEY, CHARLES A.: Pvt. B. Chesterfield Co. See 16th Va. Inf. Enl. 7/28/61 at Sewell's Point. Absent sick 5/8/62. Present (1/1/63). Age 24, 5'9", fair comp., blue eyes, brown hair, trader.

TRAYLOR, JOSEPH: Pvt. See 16th Va. Inf. Enl. 5/1/61 at Manchester. AWOL 5/4/62. Detached to Drewry's Bluff, 8/1/62. Transferred to Captain J.B. Jones Co. Art. Per S.O. #208/5, 9/5/62.

VADEN, JAMES A.: 1st Sgt. See 16th Va. Inf. Enl. 5/1/61 at Manchester. Reduced to ranks on 10/4/61 Elected 1st Sgt. on 1/4/62. MWIA at Port Republic on 6/9/62. Died of wounds, typhoid fever, and erysipelas on 6/30/62 at University of Virginia Hosp.

VEST, AURELIUS: Pvt. See 16th Va. Inf. Enl. 5/14/61 at Sewell's Point.

WARMACK, PLEASANT F.: Pvt. See 16th Va. Inf. Enl. 5/14/61 at Sewell's Point. Absent sick 7/4/62. AWOL.

WARTHEN, JOHN W.: Pvt. See 16th Va. Inf. Enl. 5/14/61 at Sewell's Point. Furloughed 2/20/62. Present through transfer to Courtney Art.

WEBB, B. L. or S.: Pvt. See 16th Va. Inf. Enl. 4/11/62 at Huger Barracks near Norfolk as a conscript. Previous service in militia. AWOL since 4/11/62.

WEISIGER, BENJAMIN B.: 2nd Lt. See 16th Va. Inf. Enl. 5/1/61 at Manchester. Elected 2nd Lt. on 1/4/62. Relieved 7/4/62.

WEISIGER, EMMETT WASHINGTON: Captain. See 16th Va. Inf. Enl. 5/1/61 at Manchester. Promoted to Capt. in 4/62. Resigned 7/15/62.

WEISIGER, WILLIAM ROBERT: Captain. See 16th Va. Inf. Enl. 5/1/61 at Manchester. Detached from the battery and assigned as surgeon in Richmond. May have later served as surgeon, 22nd Battn. VA Inf. B. 3/3/1828 in Manchester, VA. Graduated in 1848 from Medical College of Virginia, occ. physician. Member of the masonic lodge. D. 7/28/1884.

WHITEHURST, WILLIAM H.: Pvt. See 16th Va. Inf. Enl. 5/1/61 at Manchester. AWOL 5/8/62. Pres. (11/27/62). Transferred to 39th Va. Cav.

WILLIAMS, MOSES: Pvt. See 16th Va. Inf. Enl. 5/1/61 at Manchester. AWOL 12/24/62 - 1/10/63. Transferred to Courtney Art.

YOUNG, JEREMIAH: Pvt. See 16th Va. Inf. Enl. 4/11/62 at Huger Barracks near Norfolk as a conscript. Previous service in militia. AWOL since 4/11/62. Transferred to Courtney Art.?

ZEIGLER, JOHN A.: Pvt. B. Franklin Co. See 16th Va. Inf. Enl. 4/11/62 at Huger Barracks near Norfolk as a conscript. Previous service in militia. Absent sick 6/28/62. Pres. (10/31/62). Discharged 2/21/63; age 29, 5'8", dark comp., blue eyes, light hair, farmer.

Roster, Campbell Long Island Artillery

ALMAND, BRICE M.: Pvt. Enl. on 3/15/62 in Campbell Co. Transferred to the Orange Art. on 10/4/62. Paroled at Appomattox C.H., VA on 4/12/65.

ALMAND, NAPOLEON BONAPARTE: Pvt. Enl. on 3/15/62 in Campbell Co. Transferred to Orange Art. on 10/4/62. Paroled at Appomattox C.H. on 4/12/65.

ARNOLD, J. J.: Pvt. Enl. on 3/15/62 in Campbell Co. At Chimbo. #5 with fever 6/19/62 - 9/8/62, 40 day furlough granted. Transferred to Orange Art. on 10/4/62. Never joined. DFR.

ARNOLD, WILLIAM H.: Pvt. Enl. on 3/15/62 in Campbell Co. Transferred to Orange Art. on 10/4/62. Taken POW at Spotsylvania C.H., VA on 5/12/64. Sent to Fort Delaware where held until he died of chronic diarrhea on 8/8/64. Buried in the Finn's Point National Cem., Salem, NJ. Widow, Lucinda filed pension application under act of 1888 in Pittsylvania Co., VA.

ASHER, J. F.: Pvt. Enl. 3/15/62, Campbell Co. Present though transfer to Montgomery's Battery.

BARNES, JAMES F.: QM Sgt. Enl. on 3/15/62 in Campbell Co. Transferred to Orange Art. on 10/4/62. Appointed QM Sgt. of H. P. Jones' Arty. Battn. Discharged on 3/1/63, provided Lewis Frank as substitute.

BATEMAN, JAMES E.: Pvt. Enl. 3/15/62, Campbell Co. Transferred to Montgomery's Battery.

BOOKER, JOHN: Pvt. Enl. 3/15/62. Never mustered.

CARPENTER, JOHN: PWR of Morris Artillery.

CARRVILLES, JACOB: Pvt. B. Campbell Co. Enl. on 3/15/62 in Campbell Co.; age 42, florid comp., grey eyes, dark hair. According to a record discharged 4/17/62. Transferred to Orange Art. on 10/4/62. Deserted at Camp Millford on 2/1/63. Arrested at Somerville on 3/9/63 by U.S. Forces, sent to Atheneum Prison, Wheeling and then to Camp Chase, OH. Released on 3/27/63 as a Confederate deserter. Described as 5'10", B. 1838, fair comp., blue eyes, brown hair.

CARRVILLES, SAMUEL: Pvt. Enl. 3/15/62, Campbell Co. D. and claim filed 3/12/63. Settlement in cert.# 4734, $121.33.

CARTER, GEORGE WASHINGTON: Pvt. Enl. on 3/15/62 in Campbell Co. Transferred to Orange Art. on 10/9/62. WIA at Chancellorsville on 5/3/63. RTD on 8/26/63. In arrest 12/6-2/29/64. Court-martialed for mutiny on 3/7/64. Pres. thru 2/23/65. Taken POW on 3/8/65 at New Market Bridge, sent to Point Lookout, MD where held until released on oath on 6/24/65. Res. of Nelson Co., 5'7-", fair comp., blue eyes, black hair. B. 1840.

CLARK, PATRICK HENRY: Captain. D. 7/26/1862

CONNELLEY, WILLIAM J.: Sgt. B. Mecklenburg Co. Enl. 3/15/62, Campbell C.H. Discharged due to heart disease, 7/31/62; age 24, 5'11½", fair comp., dark eyes and hair, farmer.

CRADDOCK, JOHN T.: Pvt. Enl. on 3/15/62 in Campbell Co. G.H. #13, Richmond w/chronic diarrhea, 8/18/62. Sent to Mayo Island, 8/30/62. Transferred to Orange Art. on 10/9/62. In Danville Hosp. with rheumatism 11/62. AWOL 1-3/63. Detailed as teamster until 9/28/63. KIA on 5/12/64 at Spotsylvania C.H., VA.

CRADDOCK, JOSEPH W.: Pvt. B. Pittsylvania Co. Enl. 3/15/62, Ebenezer Church, Campbell Co.; age 28, carpenter/farmer, 6' 1", florid comp., blue eyes. Discharged due to cured spine, 4/7/62. Campbell p.l.

EAGLES, WILLIAM H. H.: Pvt. B. Charlotte Co. Enl. 3/15/62, Eb. Ch., Campbell Co. Discharged 4/7/62; age 39, 6' 2", light comp., grey eyes, auburn hair, farmer. Campbell p.l.

EAST, JOHN O.: Pvt. Enl. 3/15/62. Enl. Served for 2 days. NFR.

ELDER, A.C.: Pvt. Enl. 3/15/62. Transferred To Orange Artillery.

FOSTER, A. C.: Pvt. Enl. on 3/15/62 in Campbell Co. Transferred to Orange Art. on 10/9/62. Never reported.

FOSTER, DAVID F.: Pvt. Enl. on 3/15/62 in Campbell Co. Transferred to Orange Art. on 10/9/62. In Chimborazo Hosp. 10/26/63 with chronic diarrhea. RTD on 3/31/64. Taken POW on 5/12/64 at Spotsylvania C.H., VA, sent to Fort Delaware on 5/21/64. Exchanged on 9/28/64, sent to Chimborazo Hosp., furloughed for 40 days. Res. of Brook Neal, Campbell Co.

FRANCIS, WILSON: Pvt. Enl. 3/15/62 in Campbell Co., age 43. Transferred to the Morris Art. on 10/4/62. Taken POW at Spotsylvania C.H. on 5/12/64, sent to Ft. Delaware, where held until he died of acute dysentery on 12/31/64. Buried at Finn's Point, NJ.

GILCHRIST, CHARLES: Pvt. B. Campbell Co. Enl. 3/15/62. Discharged 9/4/62 due to phthisis; age 45, 5'8½", light comp., blue eyes, sandy hair, farmer.

GRISHAW, THOMAS: Pvt. Enl. 3/15/62. Died 5/15/62, Camp Lee. Cert. #23595, $37.80 due.

HAMLETE, ASA D.: Pvt. Enl. 3/15/62. D. 5/12/62, Richmond.

HAMLETE, LEE B.: Pvt. Enl. 3/15/62. D. 5/62, Camp Winder.

HANCOCK, N. M. : 2nd Lt. Enl. ? Assigned to "4th Battalion per S.O. #242/6, 10/12/63. Paroled "4th Battalion", at Appomattox.

HARPER, JOHN: Pvt. Enl. on 3/15/62 in Campbell Co., age 40. Transferred to the Morris Art. on 10/4/62. Detailed as a teamster most of the war. Paroled at Appomattox C.H., VA on 4/12/65. Res. of Hat Creek, Campbell Co., VA, died there in 1875.

HARPER, JOSEPH: Pvt. Enl. on 3/15/62 in Campbell Co., age 40. Transferred to the Morris Art. on 10/4/62. Detailed as a blacksmith. Taken POW on 3/18/65 at Waynesboro, VA, escaped, WIA. WIA in right lung at Five Forks, VA on 4/6/65. Paroled at Appomattox C.H., on 4/12/65. Postwar Farmer, res. of Morris Church in 1885.

HAYTT, Z. M.: Pvt. Son of William P. Haytt. Enl. 3/15/62. At Chimbo. #5 6/19/62. D. 7/8/62 of typhoid fever.

HOGANS, MEREDITH L.: Pvt. Enl. on 3/15/62 in Campbell Co. In Richmond Hosp with debility 8/18- 30/62. Transferred to Orange Art. on 10/9/62. AWOL 3/63. Pres. thru 4/65. Paroled on 5/30/65 at Campbell C.H., VA on 5/30/65. Res. of Mt. Airy, Pittsylvania Co.

HOLLAND, HENRY: Pvt. Enl. 3/28/62, Richmond. Transferred To the Orange Art.

HOLT, BRANCH W.: Pvt. Enl. on 3/15/62 in Campbell Co., age 35. Transferred to Morris Art. on 10/4/62. Paroled at Appomattox C.H., VA on 4/12/65. Res. of Halifax Co. Res. of Rabat in 1885.

HOLT, CABELL R.: Pvt. Enl. 3/15/62. D. 5/13/62 at General Hospital.

HOLT, CALVIN HILL: Pvt. Enl. on 3/15/62 at Campbell C.H. D. 8/24/62 at Gen. Hosp. #13, Richmond. B. 1819. Occ. farmer.

HOLT, CHARLES C.: Pvt. Enl. on 3/15/62 in Campbell Co. In hosp. 8/18-26/62 with fever. Transferred to Orange Art. on 10/9/62. Paroled at Campbell C.H., VA on 5/24/65. Res. of Hat Creek, Campbell Co., VA.

HOLT, HENRY L.: Pvt. Enl. on 3/15/62 in Campbell Co., age 38. Transferred to Morris Art. on 10/4/62. Taken POW at Spotsylvania C.H., VA on 5/12/64, sent to Fort Delaware where held until he died on 11/3/64 of an inflamation of the lungs. Buried in the Finn's Point Cem., NJ. Res. of Hat Creek, Campbell Co. "Left widow and 3 orphans."

HOLT, JAMES H.: Pvt. B. Campbell Co. Enl. 3/1/62, Campbell Co. At Chimbo. #13. At Brook Neal, Campbell Co. With Ascites. Furloughed 8/26/62 for 50 days. Discharged due to dropsy as a result of the enlargement of the liver and chronic bronchitis, 8/27/62, Richmond; age 38, 6' 1", laborer. Died and claim filed by widow, Sarah J. Holt.

HOLT, JOHN P.: Pvt. Enl. on 3/15/62 in Campbell Co. Transferred to the Morris Art. on 10/4/62. MWIA at Gettysburg, PA on 7//1/63, died there on 8/4/63, after amputation of left leg at the thigh.

HOLT, MARCELLUS P.: Pvt. Enl. on 3/15/62 in Campbell Co., age 20. Transferred to the Morris Art. on 10/4/62. WIA on 7/1/63 at Gettysburg, PA. Paroled at Appomattox C.H., VA on 4/12/65. Res. of Aspinwall, Campbell Co., VA in 1885.

HOLT, ROBERT D.: Pvt. Enl. on 3/15/62 at Campbell C.H. Died on 7/29/62 of Remittent fever at Chimborazo Hosp. #5. B. 1827. Occ. Blacksmith.

HOLT, ROBERT S.: Pvt. Enl. on 3/15/62 in Campbell Co. Transferred to the Morris Art. on 10/4/62. MWIA at Sommerville Ford on 9/14/63. Died at Chimborazo Hosp. on 10/1/63 after amputation of right leg.

HOLT, THOMAS H.: Pvt. Enl. on 3/15/62 in Campbell Co. Transferred to Orange Art. on 10/9/62. MWIA on 9/14/63 at Somerville Ford, VA, shell through both thighs, and gunshot wound in the arm. Died in General Hosp. #1, Richmond on 9/22/63. Buried in Hollywood Cem., Richmond, VA.

HUGHES, WILLIAM A.: Pvt. Enl. on 3/3/62 in Prince Edward Co. Transferred to Long Island Art. Transferred to Orange Art. on 10/9/62. Hospitalized in Farmville on 4/7/63 with scorbitus. Detailed to QM Dept. at Petersburg. B. 1811. Res. of Prince Edward Co., dark comp., grey eyes, gray hair. Occ. teacher.

HUNTER, J.W.: Pvt. Enl. 3/15/62. AWOL for 11 days at Winder Hospital 10/64. Remarks state Greenville, S.C.

HUNTER, WILLIAM J.: Pvt. Enl. 3/15/62. AWOL 11 days. Transferred To the Orange Artillery. Campbell p.l.

JOHNSTON, WILLIAM H.: Pvt. Enl. on 3/15/62 in Campbell Co. Wounded at Seven Pines. Transferred to the Morris Art. on 10/4/62. WIA, Gettysburg. Paroled at Appomattox C.H., VA on 4/12/65. In Soldiers Home, Richmond, 11/3/24, age 83.

JONES, MARSHALL B.: Pvt. Enl. 3/15/62. Transferred To Morris Art.

JORDAN, CHRISTOPHER COLUMBUS: Sgt. Enl. on 3/15/62 in Campbell Co. Transferred to Morris Art. on 10/4/62. Reduced to Cpl. on 10/5/62. KIA at Fredericksburg, VA on 12/13/62.

KABLER, JOHN H.: Pvt. Enl. on 3/15/62 in Campbell Co., age 20. Transferred to the Morris Art. on 10/4/62. Promoted to Cpl. in 1-2/63. WIA at Gettysburg, PA 7/1/63. RTD by 2/29/64. WIA at Sailor's Creek, VA on 4/7/65. Paroled at Appomattox C.H. on 4/12/65. Res. of Pigeon Run, Campbell Co. Postwar minister, moved to Bertie Co., NC.

KABLER, LILBOURN: Pvt. Enl. on 3/15/62 in Campbell Co. Transferred to Orange Art. on 10/9/62. In Hosp. on 8/22/63. DFR on 3/1/65. Paroled at Campbell C.H. on 5/29/65. Campbell Co. (Pigeon Run) p.l.

KABLER, THOMAS N.: 1st Cpl. Enl. on 3/15/62 in Campbell Co. Transferred to Orange Art. on 10/9/62. Taken POW at Spotsylvania C.H., VA on 5/12/64. Exchanged. Taken POW again at Strausburg, VA on 10/19/64, sent to Point Lookout, MD, exchanged on 2/18/65. Paroled at Campbell C.H. on 5/29/65.

KERR, GEORGE: Pvt. Enl. on 8/11/62 in Campbell Co. as a conscript. Transferred to the Morris Art. on 10/4/62. Discharged on 5/27/63. Previously served in Co. B, 11th VA Inf. B. 1833 in Campbell Co., 6'1", dark comp., dark hair, grey eyes, occ. farmer.

KERSEY, WILLIAM T.: Pvt. B. Charlotte Co. Enl. 3/15/62. Discharged due to tuberculosis; age 35, 5' 9", light comp., blue eyes, farmer.

LABEY, JOHN H.: Medical reference only.

LANE, WILLIAM: Pvt. Enl. on 3/15/62 in Campbell Co. Transferred to Orange Art. on 10/9/62. Taken POW at Spotsylvania C.H., VA on 5/12/64. Sent to Fort Delaware where held until released on oath on 5/11/65. Described as res. of Charlotte Co., VA, 5'10", dark comp., dark hair, blue eyes.

LAWSON, JOHN R.: Postwar reference. Campbell p.l.

LAYNE, J.W.: Pvt. Enl. 3/15. Transferred To Orange Art.

LAYNE, RICHARD L.: Pvt. Enl. on 3/15/62 in Campbell Co., age 18. Transferred to Morris Art. on 10/4/62. Paroled at Appomattox C.H., VA on 4/12/65. Antebellum res. of Pigeon Run, Campbell Co. Postwar res. of Roanoke Co., VA, occ. carpenter.

LIPSCOMB, LEONARD WILLIAM: Pvt. Enl. on 3/15/62 in Campbell Co., age 22. Transferred to Morris Art. on 10/4/62. WIA on 7/1/63 at Gettysburg, PA. Paroled at Appomattox C.H., VA on 4/12/65. Antebellum res. of Hat Creek. Postwar farmer at Cole's Ferry in 1885. Charlotte Co. p.l.

LOVELACE, WILLIAM D.: Pvt. Enl. on 3/15/62 in Campbell Co., age 22. Transferred to the Morris Art. on 10/4/62. Taken POW at Five Forks, VA on 4/1/65, sent to Newport News, VA where held until 7/1/65. Antebellum res. of Pittsylvania Co., VA. Living in Palmersville, TN in 1885.

MANN, JOHN F.: Pvt. Enl. on 3/15/62 in Campbell Co. Wounded. Transferred to Morris Art. on 10/4/62. Paroled at Appomattox C.H., VA on 4/12/65.

MACON, W. A.: Sgt. Postwar reference. Campbell p.l.

MASON, CLIFTON B.: Pvt. Enl. 3/15/62. Transferred To Orange Art.

MASON, JAMES H.: Pvt. Enl. on 3/15/62 in Campbell Co. Transferred to Morris Art. on 10/4/62. MWIA and taken POW at Spotsylvania C.H., VA on 5/12/64, sent to Lincoln Gen. Hosp., Washington, DC, died there on 5/19/64. B. 1839, single, res. of Brookneal, Campbell Co. Bro. of William A. Mason.

MASON, WILLIAM A.: Sgt. Enl. on 3/15/62 in Campbell Co., age 31. Transferred to Morris Art. on 10/4/62. Assigned as QM Sgt. of Morris Art. Paroled in Campbell Co. on 5/30/65, farmer. Living at Pigeon Run, Campbell Co. in 1885. Bro. of James A. Mason.

McCRAW, BOOKER: Pvt. Enl. on 8/1/62 in Campbell Co. Transferred to Orange Art. on 10/9/62. In arrest 11/4/63 thru 7/7/64 in Ward's Penal Battalion, Co. B. RTD on 8/1/64. Taken POW at Amelia C.H. on 4/6/65, sent to Newport News, VA. NFR. Campbell p.l.

McCRAW, JOHN C.: Pvt. Enl. on 3/15/62 in Campbell Co. Transferred to Orange Art. on 10/9/62. Provided James Profitt as a substitute.

McCRAW, THOMAS G.: Commissary Sgt., Enl. on 3/15/62 in Campbell Co. Transferred to Orange Art. on 10/9/62. WIA on 9/14/62 at Somerville Ford. Sent to Charlottesville Hosp. Absent on sick furlough on 2/29/64 thru 3/1/65. B. Charlotte Co. in 1845. D. 5/20/1917 in Richmond Soldier's Home. Buried in Hollywood Cem.

McCRAW, W. A.: Pvt. Enl. on 3/15/62 in Campbell Co. Transferred to Orange Art. on 10/9/62. Died of disease in Orange Co. on 2/10/63. Mother was named Elizabeth.

MEEKS, H.: Pvt. Enl. on 3/15/62 in Campbell Co. Transferred to Orange Art. on 10/9/62. Furloughed from Richmond Hosp. on 11/17/62. NFR.

MINIX, JOHN: Pvt. Enl. on 3/15/62 in Campbell Co. Transferred to Morris Art. on 10/4/62. WIA at Gettysburg, PA on 7/1/63. Taken POW at Spotsylvania C.H. on 5/12/64, sent to Fort Delaware where held until released on taking the oath of allegiance on 6/20/65. Res. of Halifax Co., 5'10", B. 2/14/1829 (1832) in Halifax Co., gray eyes, light hair, ruddy comp. Living at Nanuna, Halifax Co., VA in 1885. D. 4/7/1910 at Gladys, Halifax Co., VA.

MURRELL, CYRUS L.: Pvt. Enl. on 3/15/62 in Campbell Co. Transferred to Morris Art. on 10/4/62. Taken POW at Strausburg, VA on 10/19/64, sent to Point Lookout, MD, where he was held until released on taking the oath of allegiance on 6/15/65, dark comp., black hair, hazel eyes, 5'9".

MURRELL, D. BLAIR: Pvt. Enl. on 3/15/62 in Campbell Co. Transferred to Morris Art. on 10/4/62. Taken POW while detailed as a nurse at Jackson Hosp., Richmond, VA on 4/3/65. Paroled there on 4/18/65.

NOELL, JOSEPH R.: 4th Sgt. Enl. on 3/15/62 in Campbell Co. Transferred to Morris Art. on 10/4/62. WIA in right leg at Gettysburg on 7/1/63. Detailed to QM department. Taken POW at Amelia C.H., VA on 4/6/65, sent to Newport News, VA, where held until released on oath on 6/14/65. Res. of Charlotte Co., fair comp., dark hair, blue eyes, 5'10". Postwar farmer. In Richmond Soldiers Home 10/27/1914-1/13/1915 when released at his own request.

OLIVER, JOHN ALONZO H.: 3rd Cpl. Enl. on 3/15/62 in Campbell Co. AWOL 13 days 8/1/62. Transferred to Morris Art. on 10/4/62. WIA at Gettysburg, PA on 7/1/63. Detailed as stable manager, Lynchburg, VA. Paroled at Campbell C.H., VA on 5/30/65. Res. of Campbell Co.

PAGE, WILLIAM B.: Pvt. Enl.? Paroled 4/24/65, Winchester.

PAYNE, PHILIP S.: Pvt. Enl. on 3/15/62 in Campbell Co. Transferred to Morris Art. on 10/4/62. Provided William M. Wiatt as substitute on 5/13/63. B. 1835. Farmer. Living at Air Point, VA in 1885.

PRICE, DAVID C.: Pvt. Enl. on 3/15/62 in Campbell Co. age 21. Transferred to Morris Art. on 10/4/62. KIA by a sharpshooter in line of battle near Spotsylvania C.H. on 6/5/64.

PRICE, PARHAM F.: Pvt. Enl. on 3/15/62 in Campbell Co. Transferred to Morris Art. on 10/4/62. WIA at Chancellorsville, VA on 5/3/63, never recovered. Paroled at Campbell C.H. on 5/29/65. B. 1819. Postwar farmer. Living at Brookneal, Campbell Co., VA in 1885.

PRICE, ROBERT W.: Pvt. Enl. on 3/15/62 in Campbell Co. Furloughed on 4/10/62 for consumption. AWOL for 13 days preceding 6/1/62. Transferred to Morris Art. on 10/4/62. Detailed to Lynchburg Hosp. most of the war. B. 1841. Postwar farmer. Living at Brookneal, Campbell Co., VA in 1885.

PRICE, WILLIAM M.: Pvt. Enl. on 3/15/62 in Campbell Co. AWOL 13 days ending 6/1/62. Transferred to Morris Art. on 10/4/62. Taken POW at Spotsylvania C.H. on 5/12/64, sent to Fort Delaware where held until he died on 6/23/65 of chronic diarrhea, buried in the Finn's Point National Cem., Salem, NJ. B. 1834. Res. of Campbell Co.

PUGH, DAVID: Pvt. Enl. 3/15/62. Died?

PUGH, JAMES H.: Pvt. Enl. on 3/15/62 in Campbell Co. At Chimbo. #13 w/fever, 8/18/62. Furloughed for 50 days to home, 9/16/62. AWOL when Transferred to Orange Art. on 10/9/62. In Richmond Hosp. on 7/12/64 with typhoid and dysentery. RTD in 10/64. Pres. thru 3/1/65. Paroled at Appomattox C.H., VA on 4/12/65. Campbell p.l.

PUGH, JOHN E.: Pvt. Enl. on 3/15/62 in Campbell Co. Transferred to the Morris Art. on 10/4/62. Taken POW at Spotsylvania C.H. on 5/12/64, sent to Fort Delaware, where held until released on oath on 6/20/65. Res. of Campbell Co., dark comp., dark hair, blue eyes, 5'10". B. 1836. occ. wheelwright. Res. of Maryville in 1885.

PUGH, MARSHALL M.: Pvt. Enl. on 3/15/62 at Maryville, Campbell Co. Sick furlough on 4/10/62. AWOL for 13 days before 6/1/62. Transferred to Morris Art. on 10/4/62. Detailed thru early 1865. Paroled at Campbell C.H., VA on 5/29/65. B. 1830, occ. carpenter. Res. of Marysville, VA.

RAMSEY, DANIEL B.: Pvt. Enl. on 3/15/62 in Campbell Co. Absent sick on 4/10/62. AWOL 8 days before 6/1/62. Transferred to Morris Art. on 10/4/62. Detailed as a blacksmith. Paroled at Appomattox C.H., VA on 4/12/65. B. 1834, farmer, res. of Madisonville, VA in 1885.

RAMSEY, WILLIAM A.: 4th Cpl. Enl. 3/15/62. Transferred To Morris Art.

REAMS, JOHN H.: Pvt. Enl. on 8/11/62 in Campbell Co. Transferred to Morris Art. on 10/4/62. Promoted to Artificer in 2/63. Taken POW at Strasburg, VA on 10/19/64, sent to Point Lookout, MD where held until released on oath on 6/19/65. Dark comp., black hair, blue eyes, res. of Campbell Co. Living in Aspinwall, VA in 1885.

ROACH, WILLIAM B.: Pvt. Enl. on 3/15/62 in Campbell Co. Transferred to Morris Art. on 10/4/62. Promoted to Artificer on 10/5/62. Reduced to Pvt. by 12/31/62. Taken POW at Spotsylvania C.H. on 5/12/64, sent to Fort Delaware, where held until released on oath on 6/21/65. Res. of Charlottesville, light comp., light hair, blue eyes. B. 1844. Farmer. Living in Rough Creek, VA in 1885.

ROARK, JAMES T.: Pvt. Enl. on 3/15/62 in Campbell Co. Transferred to Morris Art. on 10/4/62. Died in camp near Frederick Hall on 12/22/63. Buried in camp. B. 1844.

ROBERTSON, J.J.: Pvt. Enl. 3/15/62. Transferred To Morris Art. Campbell p.l.

SCOTT, J. A.: 2nd Sgt. Enl. 3/15/62. Assignment per S.O.# 242/6, 10/12/63.

SCOTT, JAMES J.: Pvt. Enl. on 3/15/62 in Campbell Co. In Chimborazo Hosp. 6/3/62 with intermittent fever, sick furlough for 20 days to Campbell Co. Transferred to the Morris Art. on 10/4/62. WIA in the shoulder at Chancellorsville, VA on 5/3/63. Taken POW at Spotsylvania C.H. on 5/12/64, sent to Fort Delaware. Released on oath on 6/21/65, dark comp., dark eyes, dark hair. Res. of Marysville, Campbell Co., 5'8". B. 1844. Postwar wheelwright. Res. of Mt. Airy, VA in 1885.

SCOTT, THOMAS L.: Pvt. Enl. on 3/15/62 in Campbell Co. Transferred to Morris Art. on 10/4/62. Taken POW on 7/5/63, sent to Fort McHenry and then to Fort Delaware, and finally to Point Lookout. Exchanged on 2/18/65. Living at Marysville in 1885. Campbell p.l.

SMITH, MARTIN P.: Pvt. Enl. on 3/15/62 in Campbell Co. Transferred to Morris Art. on 10/4/62. In Winder Hosp., Richmond 5/12-14/63 with typhoid fever. In Charlottesville Gen. Hosp. 9/18/63 with debilities. Pres. thru at least 10/31/64. B. 1832. Postwar farmer. Res. of Brookneal in 1885.

SMITH, PLEASANT: Pvt. Enl. on 3/15/62 in Campbell Co. AWOL 8 days ending 6/1/62. In Richmond Hosp. 8/18-27/62 with fever. Transferred to Morris Art. on 10/4/62. AWOL in Orange Co. 7/31/63. In arrest 3-10/64. B. 1829. Res. of Charlotte Co. during the war. Res. of Baileysville in 1885. Occ. carpenter. Charlotte Co. P.I.

SMITH, S. W.: Postwar reference. Charlotte Co. p.l.

SMITH, WALTER C.: Pvt. Enl. on 3/15/62 in Campbell Co. On sick furlough 7/12/62-10/31/64. Transferred to Morris Art. on 10/4/64. Detailed as a nurse at Pratt Hosp., Lynchburg, VA 1/6-10/31/64. Taken POW at Farmville on 4/6/656, sent to Newport News. Released on oath on 7/4/65. B. 1843. Postwar merchant, Res. of Baileysville in 1885.

SPENCER, THOMAS P.: Jr. 1st Lt. Transferred To Morris Art. Assigned per S.O.# 242/6, 10/12/63.

St. JOHN, ISAAC R.: Pvt. Enl. on 3/15/62 in Campbell Co. Transferred to Morris Art. on 10/4/62. Died in Chimborazo Hosp. in late 1864. Buried in Richmond, VA. B. 1828, farmer, res. of Baileysville, VA.

St. JOHN, ISAAC W.: Pvt. Enl. on 3/15/62 in Campbell Co. Transferred to Morris Art. on 10/4/62. Paroled at Appomattox C.H., VA on 4/12/65. Res. of Rough Creek, Campbell Co. B. 1835. Living at Rough Creek in 1885.

St. JOHN, JACOB C.: Pvt. Enl. on 3/15/62 in Campbell Co. Transferred to Morris Art. on 10/4/62. Paroled at Appomattox C.H., VA on 4/12/65. B. 1830. Postwar res. of Aspinwall in 1885.

St. JOHN, JAMES A.: Pvt. Enl. on 3/15/62 in Campbell Co. WIA in side during Seven Days Battles at Fair Oaks. Transferred to Morris Art. on 10/4/62. In various hospitals most of the remainder of the war. Taken POW on 3/4/65 at Rock Fish Gap, escaped and went home. B. 1828, farmer, living in Red House in 1885.

St. JOHN, RICHARD T.: Pvt. Enl. on 3/15/62 in Campbell Co. Transferred to Morris Art. on 10/4/62. MWIA at Gettysburg, PA on 7/1/63. Died on 9/4/63 at Staunton Gen. Hosp.

St. JOHN, WILLIAM L.: Pvt. Enl. on 3/15/62 in Campbell Co. At Farmville G.H. with debility 5/28/62. Deserted 8/10/62. Transferred to Morris Art. on 10/4/62. WIA at Chancellorsville, VA on 5/3/63. B. 1822. Living at Rough Creek in 1885.

STENETTE, E.: Pvt. Enl. on 8/10/62 in Campbell Co. KIA at Chancellorsville, VA on 5/3/63. Born in Bedford Co.

STEPP, EDWARD R.: Pvt. Enl. on 8/11/62 in Campbell Co. MIA at Sharpsburg on 9/17/62. Transferred to Morris Art. on 10/4/62. Eventually RTD. Taken POW at Spotsylvania C.H., VA on 5/12/64, sent to Fort Delaware where held until released on oath on 6/21/65. Res. of Campbell Co., dark comp., dark hair, dark eyes, 5'9". B. 1831. Res. of Mt. Zion in 1885.

STORER, JOHN: Pvt. Enl. 3/15/62. At Chimbo. #5 w/typhoid, 6/18/62. D. 7/5/62.

STORER, RICHARD E.: Pvt. Enl. on 3/15/62 in Campbell Co. In Chimborazo Hosp. on 6/16-17/62 with icterus and debility. In Lynchburg Hosp. 6/17/62. Transferred to Morris Art. on 10/4/62. WIA at Gettysburg, PA. WIA at Deep Bottom in fall of 1864. WIA at Five Forks on 4/6/65. Taken POW at Five Forks, sent to City Point and then to Newport News, where held until released on oath on 7/1/65. Res. of Campbell Co., fair comp., light hair, grey eyes, 5'3". B. 1827. Farmer. Living in Evington in 1885.

SUBLETT, SAMUEL M.: Pvt. Enl. on 8/11/62 in Campbell Co., age 23. Res. of Morris Church. Transferred to Morris Art. on 10/4/62. Detailed as teamster. In Chimborazo Hosp. 8/3- 17/63 with typhoid fever and white swelling. Pres. thru 10/31/64. Paroled at Ashland, VA on 4/22/65. Farmer, res. of Red House in 1885.

TAYLOR, WILLIAM M.: Pvt. Enl. on 3/15/62 in Campbell Co. Transferred to Morris Art. on 10/4/62. WIA at Gettysburg, PA on 7/1/63. RTD by 10/31/63. Paroled at Appomattox C.H., VA on 4/12/65. B. 1832. Occ. carpenter/cabinet maker. Res. of Marysville in 1885. Campbell p.l.

TERRY, J. M.: Postwar reference. Campbell p.l.

TRENT, JOHN H.: Pvt. Enl. 3/15/62. D. 4/28/62, Camp Lee.

WALLER, PRESTON J.: Pvt. Enl. on 3/15/62 in Campbell Co. Transferred to Morris Art. Pres. thru 10/31/64. Moved to Kentucky postwar.

WALTON, JOSEPH W.: Pvt. Enl. on 3/15/62 in Campbell Co. Transferred to Morris Art. on 10/4/62. MWIA at Cedar Creek on 10/19/64, died in Mt. Jackson Gen. Hosp. on 12/16/64. Res. of Campbell Co.

WATKINS, ROBERT W.: Pvt. Enl. on 8/1/62 in Campbell Co. Transferred to Morris Art. on 10/4/62. Died of pneumonia in Chimborazo Hosp. on 6/1/64.

WILBURN, THOMAS: Pvt. Enl. 3/15/62. D. 5/12/62 at General Hospital.

WILLIAMSON, ELBERT MADISON: 1st Sgt. B. 1835. Enl. on 3/15/62 in Campbell Co. Transferred to Morris Art. on 10/4/62. Detailed to clerk for Major H. P. Jones. Wounded at Petersburg. Paroled at Farmville, VA 4/11-21/65. Described as 5'7", grey eyes, florid comp., gray hair. Postwar tobacconist. Res. of Charlotte Co. Living in Danville in 1885. D. 1935.

WIMBISH, ABRAM: Captain. B. Isle of Wight Co.? Assigned after disbandment of the battery per S.O. #242/6, 10/12/63. Paroled 4/25/65.

WOOD, JAMES B.: 2nd Cpl. Enl. ?. AWOL 13 days ending 6/1/62. Transferred to Morris Art. on 10/4/62. Detailed as a teamster from 8/63-10/64. Paroled at Campbell C.H., VA on 6/1/65.

WOODALL, JAMES G.: Pvt. Enl. 3/15/62. Died in service?

WOODALL, J. H.: Pvt. Enl. 3/15/62. D. 5/7/62, Camp Winder.

WOODALL, T. G.: Pvt. Enl. 3/15/62. D. 5/5/62 at general hospital.

WYATT, SAMUEL A.: Pvt. Enl. on 8/11/62 in Campbell Co. Transferred to Morris Art. on 10/4/62. Died of disease soon after 8/31/63 muster. Res. of Campbell Co.

Bibliography

Manuscripts

Jefferson County Museum, Charles Town, West Virginia.
Roger Preston Chew Collection

National Archives, Washington, D.C.
Compiled Service Records

Private Possession
Hess, Aaron T., Diary, Typescript
Westwood, William T., Reminiscences, Typescript

University of Virginia, Charlottesville, Virginia.
John Warwick Daniel Papers

Virginia Military Institute, Lexington, Virginia.
Alumni Records

Library of Virginia, Richmond, Virginia.
Joseph V. Bidgood Papers, Unit Record Department, Adjutant General's Office, Confederate Military Records

Published Primary Sources

Clark, Walter. *Histories of Several Regiments and Battalions from North Carolina in the Great War, 1861 - 1865*. 5 vols. Raleigh, North Carolina, 1901.

Runge, William H., ed. *Four Years in the Confederate Artillery: The Diary of Private Henry Robinson Berkeley.* Richmond, 1991.

United States War Department. *War of the Rebellion: A Compilation of the Official Records of the Union and Confederate Armies.* 128 vols. Washington, D.C., 1880-1901.

Williamson, E. M. *Confederate Reminiscences*. Danville, Virginia, 1935.

Newspapers and Periodicals

Richmond *Examiner*
Blue and Gray
Civil War Times Illustrated
Confederate Veteran, 1893-1932, 40 vols.
Daily Richmond *Enquirer*
Southern Historical Society Papers, 1876-1953, 52 vols.

Secondary Sources

Bearss, Ed. *First Manassas Battlefield Map Study*. Lynchburg, n.d.

Bohannon, Keith S. *The Giles, Alleghany and Jackson Artillery*. Lynchburg, 1990.

Bushong, Millard. *Fightin' Tom Rosser*. Shippensburg, Pa.,1983.

Carmichael, Peter S. *The Purcell, Crenshaw and Letcher Artillery*. Lynchburg, 1990.

Cavanaugh, Michael A. *The Otey, Ringold and Davidson Virginia Artillery*. Lynchburg, 1993.

Crew, R. Thomas, Jr. and Benjamin H. Trask. *Grimes' Battery, Grandy's Battery and Huger's Battery Virginia Artillery*. Lynchburg, 1995.

Driver, Robert J., Jr. *The 1st and 2nd Rockbridge Artillery*. Lynchburg, 1987.

________________. *10th Virginia Cavalry*. Lynchburg, 1992.

________________. *1st Virginia Cavalry*. Lynchburg, 1991.

Driver, Robert J., Jr. and Kevin C. Ruffner. *1st Battalion Virginia Infantry, 39th Battalion Virginia Cavalry, 24th Battalion Virginia Partisan Rangers*. Lynchburg, 1996.

Evans, Clement Anselm, ed. *Confederate Military History*. 12 vols., Atlanta, 1899.

Freeman, Douglas Southall. *Lee's Lieutenants*. 3 vols. New York, 1942-45.

Henderson, William D. *12th Virginia Infantry*. Lynchburg, 1984.

Jensen, Les. *32nd Virginia Infantry*. Lynchburg, 1990.

Johnson, Curt & Richard C. Anderson, Jr. *Artillery Hell: The Employment of Artillery at Antietam*. College Station, Texas, 1995.

Koleszar, Marilyn Brewer. *Ashland, Bedford, and Taylor Light Artillery*. Lynchburg, 1994.

Krick, Robert K. *Lee's Colonels: A Biographical Register of the Field Officers of the Army of Northern Virginia*. Dayton, Ohio, 1991.

_____________. *The Fredericksburg Artillery*. Lynchburg, 1986.

_____________. *Stonewall Jackson at Cedar Mountain*. Chapel Hill, North Carolina, 1990.

_____________. *Conquering the Valley: Stonewall Jackson at Port Republic*. New York, 1996.

Macaluso, Gregory J., *Morris, Orange, and King William Artillery*. Lynchburg, 1991.

Martin, David G. *The Fluvanna Artillery*. Lynchburg, 1992.

Moore, Robert H., II. *The Danville, Eighth Star New Market, and Dixie Artillery*. Lynchburg, 1989.

________________. *The Richmond Fayette, Hampden, Thomas, and Blount's Lynchburg Artillery*. Lynchburg, 1991.

________________. *The Charlottesville, Lee Lynchburg, and Johnson's Bedford Artillery*. Lynchburg, 1990.

Musselman, Homer D. *The Caroline Light, Parker and Stafford Light Virginia Artillery*. Lynchburg, 1992.

________________. *47th Virginia Infantry.* Lynchburg, 1991.

Sears, Stephen W. *To the Gates of Richmond: The Peninsula Campaign.* New York, 1992.

Sherwood, W. Cullen. *The Nelson Artillery: Lamkin and Rives Batteries.* Lynchburg, 1991.

________________ and Richard L. Nicholas. *Amherst Artillery, Albemarle Artillery and Sturdivant's Battery*. Lynchburg, 1996.

Tanner, Robert G. *Stonewall in the Valley*. Garden City, New York, 1976.

U.S. War Department, comp. *Atlas to Accompany the Official Records of the Union and Confederate Armies, 1861-1865*. Washington, D.C., 1891- 1895.

Virginia Military Institute. *Virginia Military Institute Register of Former Cadets*. Lexington, 1957.

Walker, Charles D. *Biographical Sketches of the Graduates and Eleves of the Virginia Military Institute.* Philadelphia, 1875.

Wallace, Lee A., Jr. *A Guide to Virginia Military Organizations*. Lynchburg, 1986.

__________. *The Richmond Howitzers*. Lynchburg, 1993.

Weaver, Jeffrey C. *Goochland Light, Goochland Turner and Mountain Artillery*. Lynchburg, 1994.

Weaver, Jeffrey C. *The Virginia Home Guards*. Lynchburg, 1996.

West, Michael. *The Gauley, Mercer and Western Artillery*. Lynchburg, 1991.

Wise, Jennings C. *Long Arm of Lee.* 2 vols. Lynchburg, 1915.